MW00724056

The Fire Safety Management Handbook

SECOND EDITION

The Fire Safety Management Handbook

SECOND EDITION

BY

DANIEL E. DELLA-GIUSTINA

College of Engineering and Mineral Resources

WEST VIRGINIA UNIVERSITY

AMERICAN SOCIETY OF SAFETY ENGINEERS ❖❖ Des Plaines, Illinois

The Library of Congress has already cataloged the paperback edition as follows:

Della-Giustina, Daniel.
The fire safety management handbook / Daniel E. Della-Giustina. -- 2nd ed.
p. cm.
Includes bibliographical references and index.
ISBN 1-885581-23-8 (pbk. : alk. paper)
1. Fire prevention--Handbooks, manuals, etc. 2. Fire prevention--Equipment and supplies--Handbooks, manuals, etc. I. Title.
TH9241.D45 1999 99-12090
628.9'22--dc21 CIP

ISBN 1-885581-41-6

Cover design: Publication Design, Inc., Allentown, Pennsylvania
Text design and composition: Cathy Lombardi, Liberty, New York
Editor: Michael F. Burditt, Manager Technical Publications, ASSE

Printed in the United States of America.

3 4 5 6 7 8 9 0

TABLE OF CONTENTS

This book is dedicated
to my three wonderful sons—
Daniel, John, and David—and
to my lovely wife Janet;
and to all the fire fighters
who lost their lives at
the World Trade Center
on 9/11/01.

ACKNOWLEDGMENTS

In preparing this text, the assistance of many individuals and organizations was utilized. Many thanks to my doctoral students in the Department of Safety and Environmental Management at West Virginia University. This book would not have been possible without the great help of our secretary, Ms. Eileen McDaniel, and the West Virginia Fire Marshal, Walter Smittle, III. A special thanks to Michael Herron and John Cragg for the immense efforts they put forth for the second revision. I would also like to thank Bryan Raughley and Anissa Hungate for their time and effort invested in the second edition. And I would like to give special thanks to Rocky Mazza for his computer skills and dedication to this edition, and to J. D. Brown, CSP, and John L. De Roo, P.E., CSP, for providing thorough technical reviews of the manuscript.

I wish to express my appreciation to the many agencies and educators who have made contributions to this book. I am indebted to many sources for materials used for which permission to reprint has been secured and proper credit given. To all these people and agencies I am grateful.

Special recognition and thanks are extended to the following:

Walter Smittle, III
West Virginia State Fire Marshal
Charleston, WV 25301

Michael W. Caravasus
Fire Fighter and Rescue Team
Morgantown Fire Department
Morgantown, WV

Dave Custar
Fire Fighter
West Virginia Fire Department
Morgantown, WV

Joseph Spiker
Extension Agent and Fire Fighter/Chief
Greensboro, PA

Ansul Inc.
Marinette, WI 54143

Industrial Risk Insurers
Hartford, CT 06102-5012

National Fire Protection Association
1 Batterymarch Park
Quincy, MA 02269

William Moser
Extension Agent
West Virginia Center for Emergency Response
Morgantown, WV

C. Everett Perkins Jr.
Division Leader
Fire Service Extension
West Virginia University
Morgantown, WV

Cerberos Pyrotronics
Cedar Knolls, NJ 07927-9985

1996 North American Emergency Response Guidebook
United States Department of Transportation
Washington, D.C. 20590-0001

United States Department of Transportation
Research and Special Programs Administration
Office of Hazardous Materials Initiatives and Training (DHM-50)
Washington, D.C. 20590-0001

Lastly, my thanks to my family for tolerance and understanding during the long hours of research and writing.

Daniel Della-Giustina, Ph.D.

PREFACE

This text presents the elements that comprise an effective fire safety management program. It was written for managers who are accountable for fire safety as part of a comprehensive safety or risk management program.

Safety programs are typically evaluated based upon the results they achieve for their respective organizations. Tangible results of any safety program can be difficult to measure. Over the years, the profession has evaluated safety program effectiveness by measuring the failures produced, such as accident frequency and severity rates, or property loss rates. Measuring safety programs by their failures is counterproductive. By the time any safety program produces the failures to measure, it is too late for managers to implement activities that could have prevented those failures from occurring in the first place.

While the safety profession has never proven that a direct correlation exists between various safety program activities and achieving favorable program results, safety managers strive to identify the possible relationships. Successful safety managers place emphasis—such as their time and organizational resources—on implementing proactive activities that impact the results of their safety programs. Safety program effectiveness should be measured by the quality, rigor, and utility of these activities, as well as their impact on the bottom line.

Having established that an effective safety program emphasizes proactive activities, this text places special attention on the fire safety activities that can achieve the most optimum results. Developing and implementing an effective fire safety management program can:

- reduce property-loss insurance premiums
- help minimize the financial impact of business interruptions
- boost customer service and public images
- foster an efficient work environment
- help realize quality gains
- impact favorably on the profitability of an organization

Special attention has been given to fire safety activities that achieve results. These activities are explained in each chapter.

Objectives of the Text

Individuals who utilize this text should be able to:

1. Identify agency assistance and available resources for fire service operations.
2. Determine organizational patterns for fire service operations.
3. Summarize qualifications expected of personnel attached to organizations providing fire services.

4. Describe the uses and operations of various types of fire control equipment.
5. Determine and identify materials considered hazardous.
6. Recognize the training and educational experiences needed for fire service personnel.
7. Ascertain the components of fire service communications and dispatching.
8. Demonstrate accepted management practices needed to establish and improve fire service operations.

The purpose of this book is to present, in an organized and sequential way, how to develop an effective fire safety management program. Numerous books and articles have been published on fire science. However, the majority thus far were concerned with the scientific aspects of fire safety as opposed to actual program management. This publication attempts to fill that gap by providing an analysis of how to manage a fire safety program, which is usually part of an overall loss control program.

The success of any organization depends on the soundness of its management system; this is no less true in the management of fire safety. Those same techniques that have been the hallmark of efficiency and profitability in the operation of any organization must be utilized in the successful management of fire safety.

A basic knowledge of the available resources and fire safety organizations is essential. This is described in Chapter 1. Knowing where to go, who to contact, what facilities and equipment they possess, and their ability to respond will be of great assistance in organizing a plan of action. In addition, it will assist management in avoiding costly duplication of equipment. Knowledge of fire fighting resources at one's command is one of the keys in determining whether a fire of a certain magnitude can be controlled with a minimum of damage, or whether it can accelerate into a major catastrophe.

The chemistry of fire is reviewed in Chapter 2. Some personnel who are involved in fire safety from the scientific aspects are not interested in the management aspects. However, for those personnel who aspire to manage fire safety applications, this chapter will provide the necessary understanding of fire chemistry essentials.

To reduce the effects and losses due to fire, Chapter 3 describes some efforts that can be used to develop an effective fire safety management program.

Chapter 4 explains the precautions and procedures that should be undertaken to identify and control hazardous materials.

Building construction is crucial for assuring life safety and controlling related fire risks. Building construction as related to fire safety is described in Chapter 5.

Chapter 6 provides an overview of commonly installed fire detection systems. Various occupancies require different types of detection systems and, in some cases, more than one type of system would be satisfactory. These systems are described in sufficient detail to allow a safety manager to make sound decisions regarding their application.

The types and functions of fire control equipment are described in Chapter 7. As with detection systems, specific conditions warrant certain types of equipment. The chapter discusses advantages and disadvantages of the various fire control equipment. This will help a safety manager assess which equipment would be the most operationally cost effective for their particular application.

Chapter 8 describes the practices that should be followed to care, maintain, and inspect fire protection systems. Particular emphasis is placed on management's responsibility to support a preventive maintenance program.

Chapter 9 explores the different types of legislation and enforcement that exist on the federal, state, and local levels, and how they are an integral part of a successful fire safety program.

With everyone now much more aware of the threats posed by acts of terrorism as well as by natural disasters, Chapter 10 was added to give the safety professional an understanding of emergency response planning.

Chapter 11 delineates the mission of the United States Fire Administration and presents a brief history of the USFA and the National Fire Academy.

CASE STUDY MODEL

In this text most chapters contain examples of case studies which illustrate chapter emphasis.

The following case study, Fire Incident Event, is a model that would encourage a group discussion by enlarging on the various levels of performance for each of the case studies. All case studies include a "Summary Description of the Event" and the "Post-Response Assessment."

CASE STUDY

Fire Incident Event

Date of Accident:	August 31, 1995
Time of Accident:	08:00 AM
Location of Accident:	Federal R&D Laboratory, Combustion Test Facility
Losses Incurred:	Major property damage, one fatality, three employees transported to hospital

Summary Description of Event

Maintenance personnel, working overhead of the combustion unit which had been shut down after testing, dropped a wrench which fell through the grating and damaged a connection on the hydrogen feed system. This occurred at the close of the shift, and potential damage was not detected. Hydrogen had been leaking throughout the night. The facility is located in the center of the site and is a two-story building that is used for combustion research. There are two combustor cell areas, a fuel handling and storage area, shop, control rooms, offices, a conference room, and data acquisition and support function areas. The building is occupied during business hours by research and support personnel.

The building is separated into two fire areas. The roof is gravel and built-up metal deck on steel beam and column support. Exterior walls are hollow concrete block with dryvit insulation on the outside. The second level exterior walls are metal panel on steel frame support. Under high-pressure stress the exterior walls in the combustion cell areas are designed to blow out before the pressure-resistant doors or windows. The interior walls of the combustion cell area are filled concrete block. The windows and doors between the control room and the combustion areas are pressure resistant and acoustical (not fire rated). There are several roll-up steel doors and the only windows are those in the exterior doors. Interior walls for the other areas are gypsum board on metal stud and extend to just above the ceiling—which is mineral tile on suspended, T-bar grid. The above-ceiling space is noncombustible. The floor is poured concrete on precast concrete slabs on masonry walls. The mezzanine level is open grid steel.

The Combustible Gas Detectors (which were located at the mezzanine floor and other locations in the facility) had not detected the hydrogen atmosphere. As personnel began their daily work just before 8 AM, the hydrogen that had accumulated toward the ceiling, ignited when the high-intensity lights illuminating the work area were turned on. The exterior walls in the mezzanine area were blown out. The sprinkler system became inoperable due to the explosion; however, water was cascading from the cooling tower connection used to quench exhaust fumes from the combustor. Fire was observed in various areas in the facility and was fed by natural gas leaking from lines damaged by the initial explosion. All power to the building was lost due to the explosion.

Debris from the building fell outward and caused carbon monoxide and ammonia gas to be released from a cylinder bank on the south side of the building. There were additional hazardous materials inside the building and the integrity of their containers was unknown.

One person walking near the area collapsed from carbon monoxide and ammonia gas exposure. The local fire department arrived at the scene about 8:15 AM, and a recovery team was then sent in to retrieve the employee who had been exposed to the carbon monoxide and ammonia. The site was secured by the fire department at approximately 9:15 PM, and the entry team found three additional victims. Two were alive but unconscious, and the medical team recovered the two unconscious victims for decontamination. Both were then transported via ambulance to the local hospital. The third victim, found in the mezzanine, was pronounced dead on the scene.

Post-Response Assessment

The Safety and Health Team, together with the local fire marshal, assembled for the incident investigation, concluded that failure to follow procedure contributed to a set of conditions that resulted in loss of life and significant property damage. They determined that, while the structure design was compliant with building code requirements and performed according to expectations, established procedures were not followed.

Maintenance workers, working over the top of the combustion unit, neglected to investigate thoroughly the potential damage to objects below their work area. Since it was an open grating on which they were working, adequate fall protection was not utilized to minimize tools or objects from falling to the lower levels. When the wrench fell to the lower levels, the workmen did not adequately investigate the extent of the damage or assess whether there was a problem that needed to be immediately addressed. Investigation of the centralized monitoring system for the area where the combustion unit was in operation did not indicate any buildup of gases. The centralized monitoring system and the gas analyzers are equipped with self-check and calibration, and continuous records indicated no malfunctions. A contributing cause to the buildup of explosive gases in the mezzanine area was an inadequate number, and inappropriate location, of detector heads.

Examination of the overhead high-intensity lamps indicated the lights were a potential ignition source. While not conclusive, when the main overhead lamps were turned on, an arc could occur causing the accumulated gas to explode.

A fatality resulted from the explosion. The individual who was killed was calibrating equipment in a gas analysis room adjacent to the combustor, prior to the start of the work shift. The two injuries in the building resulted when the other two workers arrived to begin preparations for the day's testing. Their first action was to increase the overall illumination in the work area. When they turned on the lights, an arc occurred and the gas exploded. All external feeds, including electricity and gas that fed the building, could be controlled exterior to the building. All utilities were shut off very early in the incident and minimized any further property damage or personnel injuries.

Recommendations

Inadequate safety analysis was conducted for the use of compressed gases. The building is designed with explosion panels, the purpose of which is to relieve the pressure in the event of a system failure on the combustion unit. While the use of these gases at that location is within the requirements for handling and storing of these materials, protection to the rack was inadequate. Had the safety factor designed into the protection of the compressed gas rack been adequate, the potential for injuries outside of the building would not have occurred.

REFERENCES

Building Officials and Code Administrators International, Inc. *BOCA National Fire Prevention Code*. Country Club Hills, IL: Building Officials and Code Administrators International, Inc., 1996.

Brauer, R. L. *Safety and Health for Engineers*. New York: Van Nostrand Reinhold, 1994.

Bugbee, P. *Principles of Fire Protection*. Quincy, MA: National Fire Protection Association, 1978.

Campbell, R. L., and Langford, R. E. *Fundamentals of Hazardous Materials Incidents*. Chelsea, MI: Lewis Publishers, 1991.

Comeau, E. *Furniture Manufacturing Facility Dust Explosion—November 20, 1994*, NFPA Fire Investigations. Quincy, MA: National Fire Protection Association.

Comeau, E., and Sawyer, S. *Dormitory Fire-Franklin MA—October 25, 1995*, NFPA Fire Investigations. Quincy, MA: National Fire Protection Association.

Cote, A. E. *Fire Protection Handbook*. Quincy, MA: National Fire Protection Association, 1986.

Della-Giustina, D. E. *Safety and Environmental Management*. New York: Van Nostrand Reinhold, 1996.

Drysdale, D. *An Introduction to Fire Dynamics*. New York: John Wiley & Sons, 1996.

Fire Protection Qualification Standard. vol. 1. Washington, D.C.: U.S. Department of Energy, 1995.

Fire Protection Qualification Standard. vol. II. Washington, D.C.: U.S. Department of Energy, 1995.

Isner, M. S., and Smeby, Jr., L. C. *Bricelyn Street Fire, Multiple Fire Fighter Fatality, Pittsburgh, PA, February 14, 1995*, NFPA Fire Investigations. Quincy, MA: National Fire Protection Association.

Lathrop, J. K. *Life Safety Code Handbook*. Quincy, MA: National Fire Protection Association, 1991.

National Fire Protection Association. *NFPA 101, Life Safety Code*. Quincy, MA: National Fire Protection Association, 1994.

Nelson, G. O. *Gas Mixtures: Preparation and Control*. Chelsea, MI: Lewis Publishers, 1992.

Pipitone, D. A. *Safe Storage of Laboratory Chemicals*. New York: John Wiley & Sons, 1991.

Puchovsky, M. T. *Automatic Sprinkler Systems Handbook*. Quincy, MA: National Fire Protection Association, 1996.

Society of Fire Protection Engineers. *Fire Protection Engineering*. Quincy, MA: National Fire Protection Association, 1995.

Fire Safety Organizations and Resources

CHAPTER 1

CHAPTER CONTENTS

This chapter provides the safety manager with a guide to obtaining pertinent information on fire safety subjects such as fire prevention, fire protection, and life safety. Its intent is to give a brief overview of the major organizations involved in the field of fire safety.

NATIONAL FIRE PROTECTION ASSOCIATION

The National Fire Protection Association (NFPA) is composed of over 73,000 members. It is a collective voice of a wide range of fire safety professionals. The NFPA's mission is to prevent the loss of life and protect property from fire. It is headquartered in Quincy, Massachusetts. The NFPA meets semiannually at its annual and fall meetings.

Organized in 1896 as an independent, nonprofit organization, the NFPA is the oldest fire safety association in the country. It maintains an open door membership policy. Any organization or individual interested in its purpose is welcomed. Its membership includes over 150 national and regional societies and approximately 32,000 individuals, corporations, and organizations. Voting members can participate in one of the following sections:

- Architects, Engineers, and Building Code Officials Section
- Aviation Section
- Education Section
- Electrical Section
- Fire Marshals Section
- Fire Science and Technology Educators Section
- Fire Service Section
- Health Care Section
- Industrial Fire Protection Section
- Lodging Section
- Rail Transportation Systems Section
- Research Section
- Wildland Fire Management Section

The NFPA is a technical and educational organization. Its major technical activity is developing, publishing, and distributing consensus standards. Their collection of consensus standards are organized into volumes called the *National Fire Codes*. The NFPA's collection of consensus standards includes widely used documents such as the *Life Safety Code®* (*NFPA 101®*), *Fire Prevention Code* (*NFPA 1*), *National Electrical Code®* (*NFPA 70*), and the *Flammable and Combustible Liquids Code* (*NFPA 30*). The standards are written such that they can be adopted into laws and regulations or incorporated by reference. Many federal, state, and local governments have adopted specific NFPA standards for enforcement. Examples include the Occupational Safety and Health Administration and state and local fire agencies. Other consensus standards have been developed for fire protection systems; fire protection of industrial operations, processes, and equipment; fire department operations and equipment; and numerous other topics. Engineering support is available to the membership for consensus standard interpretations and related assistance.

The NFPA's standards are developed through a committee consensus format. Each committee is composed of individuals who represent a balanced cross section of interests and opinions from various groups within the fire safety community. Standards are developed in an organized manner. First, the NFPA publishes a call for proposals. Any individual, group of individuals with a common interest, or organization can identify the need for a new standard or amendment to an existing standard during this proposal phase. The proposed standard or amendment is published in the NFPA's Technical Committee Reports on Proposals for public review and comment. Any individual or organization can send the NFPA comments about the proposed standard or amendment. Each comment is in turn published in the Technical Committee Reports on Comments, which gives the membership an opportunity to study and validate the comments. The committee prepares a final report for the membership. At this point, the new standard or amendment has been openly reviewed by the public. The proposed standard or amendment is then voted on by the membership at the annual or fall meeting. If the proposed standard or amendment is favored by the membership, then the NFPA Standard Council officially issues the new standard or amended existing standard.

The educational activities include seminars on consensus standards, a public education program called *Learn Not To Burn*, publishing books such as the *Fire Protection Handbook*, investigating and reporting on large loss of life and property to provide lessons learned, maintaining a fire experience database, and an extensive technical library. The NFPA provides a wide range of books, training packages, educational materials, and visual aids. Members receive two publications: *Fire News* and *Fire Journal*. Members also receive a discount on publications, seminar rates, and subscriptions to the *National Fire Codes*.

UNDERWRITERS LABORATORIES, INC.

Underwriters Laboratories, Inc. (UL) was founded in 1894. It is chartered as a not-for-profit, independent organization that provides testing services for the public. This corporation maintains and operates laboratories for testing devices, systems, and materials. UL determines whether they meet safety standards affecting life and property.

UL also conducts organized Engineering Councils to assist in establishing its own requirements. This assumes that its findings are based on adequate field experiences, as well as laboratory tests and engineering decisions. These combined groups then review reports on products prior to their release to the public. The councils represent a diversified membership of experts in specific areas. The Fire Engineering Councils functioning in UL's major areas of interest are burglary protection, casualty, building construction, electrical, fire protection, and marine.

Annual directories are published by UL and may be ordered by writing to the address listed in Appendix 3 of this book. These directories include the following:

- Automotive, Burglary Protection, and Mechanical Equipment
- Building Materials

- Roofing Materials and Systems
- Fire Protection Equipment
- Fire Resistance
- Gas & Oil Equipment
- Classified Products
- Hazardous Location Equipment
- Electrical Appliance & Utilization Equipment
- Electrical Construction Materials
- General Information for Electrical Construction, Hazardous Location, and Heating and Air Conditioning Equipment
- Marine Products
- Directory of Appliances, Equipment, Construction Materials, and Components Evaluated in Accordance with International Publications

UL STANDARDS/AMERICAN NATIONAL STANDARDS INSTITUTE (ANSI)

The UL has a standards writing body that writes standards based on the expertise of knowledgeable safety professionals who conduct the proper investigations and research prior to a standard's implementation. The knowledge base and clarification of the requirements is important in order to authorize the application of UL marks on manufactured products. UL marks address the ability to modify existing standards in the advancement of technology. The ANSI process requires that approvals shall be renewed every five years, and UL Standards are constantly reviewed, revised, and updated to the ANSI status. However, UL's goal is to obtain ANSI approval for the standards it develops. UL is accredited by ANSI to utilize different approaches to gain ANSI approval. Approximately 70% of UL's Standards are approved as American National Standards by the American National Standards Institute (ANSI).

INSURANCE COMPANIES

Insurance companies have a unique role in the fire safety community. Typically, a business customer purchases an insurance policy for a stipulated one-year period of time. During this period, an insurance company agrees to underwrite an organization's risks for a fee, often called a premium. If an organization experiences a loss such as a fire, then the insurance company agrees to indemnify the organization for its loss.

Insurance premiums charged by insurance companies are based upon the types of risks encountered in an organization's operations, as well as an organization's rigor in managing its risks. Insurance companies can give credits to organizations with properly maintained fire protection systems, written and rehearsed emergency plans, practiced fire prevention policies and precautions, and a general commitment to loss prevention.

Commercial insurers typically maintain loss prevention specialists on staff. These loss prevention specialists are trained in fire protection engineering, emergency planning, risk management, security, and business interruption planning. Their services are usually included in the cost of the premiums. Traditionally, insurers conduct regular inspections of their insureds' properties to assure that risk management practices and procedures are being implemented. Today however, insurers are taking on the role of consultants for their insureds. No longer are there days of tedious inspections that result in a punch list of recommendations. Good insurers have shifted their focus to creating partnerships with their insureds that identify and implement risk management strategies to realize mutual profits for the insured and the insurer. Wise safety managers are taking advantage of their insurers' loss prevention staffs and resources. In addition to consulting in technical areas that a safety manager might be weak in, loss prevention staffs can help safety managers measure the effectiveness of loss prevention activities in an attempt to favorably impact profitability.

FACTORY MUTUAL

Insurance companies such as the Factory Mutual System (FM) and Industrial Risk Insurers (IRI) are at the forefront of industrial fire safety. They provide engineering and inspection services, development of standards and fire research, and development of their own standards.

Founded in 1835, the Factory Mutual System consists of four member companies. These include the Allendale Mutual Insurance Company, the Arkwright-Boston Insurance Company, the Protection Mutual Insurance Company, and Philadelphia Mutual Company. The Factory Mutual System stipulates good loss control practices as a prerequisite when issuing insurance coverage to manufacturing plants.

The Factory Mutual System conducts basic and applied research, develops standards, and issues approval on materials and fire protection equipment. Inspections and evaluations are conducted by the Division of Engineering. They analyze existing hazards, the protection systems, and management's property conservation methods.

The Factory Mutual Loss Prevention Data Book Service provides information on sprinkler systems, water supplies, extinguishers, industrial hazards, construction, heating systems, and many other subjects. The data book is very informative and includes illustrations that are clear and concise.

The Factory Mutual System also offers the annual *Approval Guide*. This publication lists protective equipment, building materials, safeguards for combustion, and many more items which have been tested by the organization. This system is considered the most significant in the advancement of fire protection technology. Information is available by writing to the address listed in Appendix 3 (p. 171).

INDUSTRIAL RISK INSURERS

In December 1975 Factory Insurance Association and the Oil Insurance Association combined to form Industrial Risk Insurers. Industrial Risk Insurers requires all customers to meet the following conditions as a prerequisite for coverage:

- Industrial properties with hazards adequately protected by a sprinkler system
- Willingness and determination by management to reduce the probability of loss
- Specialized underwriting, inspection, engineering, and loss prevention service
- A premium which substantially supports the costs of these services

Industrial Risk Insurers operates a fire safety laboratory in its home office. This is essentially for training its own engineers. Another program is sponsored for its policyholders and agents in the proper use of fire protection devices.

NATIONAL FIRE ACADEMY

One of the central concerns addressed by the United States Fire Administration (USFA) was the development of a National Fire Academy (NFA). Until the NFA site was operational, an interim program of traveling mini-courses was used to teach the NFA's courses. During this time a permanent site for the NFA was selected at the former St. Joseph College in Emmitsburg, Maryland. On January 21, 1980 the NFA formally opened its doors. The campus, located on 110 acres of wooded land, has 38 fully equipped classrooms, two large auditoriums, and three dormitories. The initial enrollment of 150 students a week has since increased to 300 students a week for a total of 1200 student-weeks of training each year.

The NFA's charge under the National Fire Prevention and Control Act of 1974 is to (1) advance the professional development of fire service personnel and of other persons engaged in fire prevention and control activities; (2) encourage new programs and recommend strengthening of existing programs of education and training at state and local levels and through private institutions by providing assistance as prescribed by the Act; and (3) perform other functions as prescribed by the Act or as the Administrator shall assign.

The NFA has had a number of salutary effects upon the fire service. For example, the NFA helps fire departments reduce injuries, deaths and property losses. Individual fire departments have discovered superior techniques for coping with fires, but their successes often have not been shared with other departments. The NFA provides a medium for fire fighters and fire officers from all across the country to interact during classes and share their experiences. NFA courses in command strategy and tactics could be attuned to specific categories of risk, such as congested cities, industrial complexes, and wildlands. Also, courses in such fields as arson investigation, code enforcement, and fire safety education address major ways of reducing fire losses.

Another benefit of the NFA is that its training can prepare fire officers with the technical expertise they need in today's advanced society. Courses in management techniques can help fire officers compete for budgetary dollars with other municipal departments. Furthermore, such courses can also help them identify antiquated practices that should be abandoned, as well as help them to assess the advantages of new management practices and equipment.

A third service NFA provides is helping fire departments shift priorities toward fire prevention. NFA courses can acquaint fire officers not only with fire prevention practices that work, but also with sound record keeping methods that prove that they work.

In successfully fulfilling its mission the NFA has also increased the attractiveness of fire service careers. The training opportunities offered have made the positions in the fire services intellectually more stimulating. At the same time, officers educated by the NFA are sought far and wide. Volunteers also have benefited from training at the NFA. Often, training for volunteers at the community level is seriously neglected. The NFA can help supplement this training by making available course material and demonstration projects, by accrediting programs, and by lending special instructors to these programs.

Any person with substantial involvement in fire prevention and control, rescue, or emergency management activities is eligible to apply for NFA courses. Applicants must meet the specific selection criteria outlined with each course description in the catalog. The selection of applicants is also based upon the impact the applicant will have on the quality of fire protection in the local community, the potential utilization of the skills acquired, and an equitable and representative distribution from the total fire service.

Each applicant must complete the standard General Admission Application Form which must be endorsed by the chief official of the organization the student is representing, or by the State Training Director's office, or by the State Fire Marshal's office. A supply of admission applications and course brochures is available by writing or calling the National Emergency Training Center (NETC) Admissions Office and through the State Fire Training Directors, large metropolitan area fire chiefs, and the ten FEMA Regional Offices throughout the country. Completed applications should be sent to the Office of Admissions and Registration, National Emergency Training Center, 16825 S. Seton Avenue, Emmitsburg, Maryland 21727.

No tuition is charged for Institute courses, and all instructional materials are provided.

NATIONAL FIRE ACADEMY CURRICULUM*

Management Science Curriculum
Fire Service Communication
Organizational Theory in Practice
Interpersonal Dynamics in Fire Service Organizations

Hazardous Materials Curriculum
Hazardous Materials Operating Site Practices
Hazardous Materials Incident Management

*Courtesy, United States Fire Administration–National Fire Academy, 1999.

Fire Prevention Curriculum

Management Curriculum
Code Management
Management of Fire Prevention Programs
Strategic Analysis of Community Risk Reduction

Technical Curriculum
Fire Inspection Principles
Principles of Fire Protection: Structures and Systems

Public Education Curriculum
Presenting Effective Public Education Programs
Developing Fire and Life Safety Strategies
Community Education Leadership
Community Education Leadership: Volunteer Incentive Program

GOVERNMENT AGENCIES

The Department of Agriculture is responsible for 186.5 million acres of national forest and grassland within the United States. This land is under the jurisdiction of the U.S. Forest Service, which maintains its own division of fire control and conducts fire research at nine experimental stations and a forest products laboratory. The Forest Service's quarterly publication, *Fire Control Notes*, is available from the Superintendent of Documents, U.S. Government Printing Office, Washington, D.C. 20402.

The Department of Commerce has jurisdiction over the Fire Research and Safety Act through a division of the National Bureau of Standards (NBS), and has long concerned itself with fire protection engineering in the establishment of standards for structural building components. It is also concerned with the fire characteristics of various materials and methods used in construction. The NBS also establishes standards for interior furnishings and clothing for the protection of the public against fire hazards in these areas.

The Federal Fire Council, organized in 1930 on an informal basis, has grown to a full-time, officially recognized agency charged with the responsibility of disseminating fire safety knowledge to federal government agencies. The council was transferred to the National Bureau of Standards, which administers the Fire Research and Safety Act.

The Labor Department is also directly concerned with occupational safety. Fire prevention inspection and training has been part of its program on safety standards under the Bureau of Labor Standards and the Office of Occupational Safety. Working with the Department of Labor are the Department of Transportation and the Interstate Commerce Commission. The Department of Transportation establishes regulations for the shipment of hazardous materials by truck, rail, water, air, or pipeline.

The appendices list additional information concerning each of the given areas as well as a list of references concerning further information about fire protection and a list of test methods for protective clothing.

REFERENCES

Building Officials and Code Administrators International, Inc. *BOCA National Building Codes*. Country Club Hills, IL: Building Officials and Code Administrators International, Inc., 1995.

Colburn, Robert E. *Fire Protection and Suppression*. New York: McGraw-Hill Book Company, 1975.

"Federal Disaster Unit: Recombinant FEMA." *Science News*, December 2, 1978, pp. 390–392.

Federal Emergency Management Agency. *Annual Report—U.S. Fire Administration*. Washington, D.C.: U.S. Fire Administration.

______. *EMS Resource Exchange Bulletin*. Washington, D.C.: U.S. Fire Administration.

______. *Fire in the United States: Deaths, Injuries, Dollar Loss, and Incidents at the National, State, and Local Levels in 1978*. 2d ed. Washington, D.C.: Fire Data Center, 1982.

______. National Fire Academy. *Today's Career: Meeting Tomorrow's Challenges*. Washington, D.C.: United States Fire Administration, National Fire Academy, 1979.

______. *Resource Exchange Bulletin*. Washington, D.C.: United States Fire Administration, Office of Planning and Education.

______. *United States Fire Administration Catalog*. Washington, D.C.: United States Fire Administration, National Fire Data Center, 1995.

______. *United States Fire Administration Fellowship Program*. Washington, D.C.: United States Fire Administration, 1980.

______. *Welcome to the National Emergency Training Center*. Washington, D.C.: Federal Emergency Management Agency, October 12, 1981.

"FEMA: Consolidating U.S. Disaster Aid; an Interview with John Macy." *Nation's Cities Weekly*, January 7, 1980, p. 3.

Gratz, David B. *Fire Department Management: Scope and Method*. Beverly Hills, CA: Glencoe Press, 1974.

Kalmerus, Leo J. *A Preliminary Report on Fire Protection Research Program Fire Barriers and Suppression*. Washington, D.C.: National Technical Information Service, 1978.

National Emergency Training Center. *Emergency Management Institute: FY 1983 Resident Training Program*. Washington, D.C.: Federal Emergency Management Agency, 1983.

National Fire Academy. *Catalog of Resident Courses 1982–83*. Washington, D.C.: National Emergency Training Center, 1982.

______. *Federal Emergency Management Agency U.S. Fire Administration National Fire Academy*. Emmitsburg, MD: The National Fire Academy, 1980.

National Fire Protection Association. *Life Safety Code Handbook*. Quincy, MA: National Fire Protection Association, 1995.

______. *National Fire Code*. Quincy, MA: National Fire Protection Association, 1995.

Planer, Robert G. *Fire Loss Control*. New York: Marcel Dekker, 1979.

United States Committee on Science and Technology. *Earthquake and Fire Act Authorization*. U.S. House of Representatives, Ninety-seventh Congress. Washington, D.C.: Government Printing Office, 1981.

______. *Expressing the Sense of the Congress that State and Local Governments Should Support the Fire Safety Efforts of the U.S. Fire Administration to Reduce Lives and Property Damage Lost by Fire*. Washington, D.C.: Government Printing Office, 1982.

United States Department of Commerce. *Accreditation in Fire Training and Education: The Final Report of the Advisory Committee on Fire Training and Education of the National Academy for Fire Prevention and Control.* Washington, D.C.: Government Printing Office, 1979.

______. *The First Annual Report of the Secretary of Commerce on Implementation of the Federal Fire Prevention and Control Act of 1974.* Washington, D.C.: Government Printing Office, June 30, 1975.

______. *National Fire Academy: A Study of the Relationship of the National Fire Academy to the Fire-Related Education Programs in Colleges and Universities.* Washington, D.C.: United States Fire Administration, National Fire Academy, 1979.

United States Fire Administration. *Fifth Annual Report.* Washington, D.C.: Federal Emergency Management Agency, 1978.

______. *Fully Involved.* Washington, D.C.: Federal Emergency Management Agency, 1979.

______. *Master Planning Report to Congress: A Report Submitted to the Congress by the Federal Emergency Management Agency.* Washington, D.C.: Government Printing Office, 1981.

______. *Public Education Resource Bulletin.* Federal Emergency Management Agency.

______. *Public Fire Education Planning: A Five Step Process.* Federal Emergency Management Agency, Office of Planning and Education. Washington, D.C.: Government Printing Office, 1980.

United States House Committee on Science and Technology. *Reauthorization of the Federal Fire Prevention and Control Act.* U.S. House of Representatives, Ninety-sixth Congress, Washington, D.C.: Government Printing Office, 1979.

United States Senate Committee on Commerce, Science, and Transportation. *Nominations: Hearing before the Committee on Commerce, Science, and Transportation, United States Senate.* Washington, D.C.: Government Printing Office, 1982.

______. *Nomination and Reauthorization of the Federal Fire Prevention and Control Act.* Washington, D.C.: Government Printing Office, 1979.

United States Statutes and Laws. *Federal Fire Prevention and Control Act of 1974 Amendments.* Washington, D.C.: Government Printing Office, 1978.

STUDY GUIDE QUESTIONS

1. What organization publishes the *National Fire Codes*?
2. Describe the National Fire Protection Association's consensus standard development process?
3. What directories does Underwriters Laboratories publish?
4. What publications does Factory Mutual publish?
5. What are the prerequisites for coverage by the Industrial Risk Insurers?

Fire Chemistry

CHAPTER 2

CHAPTER CONTENTS

FIGURES

INTRODUCTION

The safety manager should have a working knowledge of basic fire science and chemistry. A fire, or combustion, is a chemical reaction. An understanding of the chemical reaction is the basis for preventing fires, as well as extinguishing fires once they initiate. A working knowledge of basic fire science and chemistry is essential for developing and implementing a successful fire safety program.

DEFINITION OF FIRE

A fire is a chemical reaction. There are many variables that can affect a fire. Effective fire safety management programs control the variables that can affect a fire. Therefore, it is imperative to understand the variables.

A fire is self-sustained oxidation of a fuel that emits heat and light (Factory Mutual Engineering Corporation, 1967). A fire requires three variables to initiate: a fuel, oxygen, and heat.

FIRE TRIANGLE

The fire triangle is a well-known representation of the three variables needed to initiate a fire. In order to initiate a fire, fuel, oxygen, and heat are required. These three variables form the fire triangle as shown in Figure 2–1 (West Virginia University Fire Service Extension).

To further understand the fire triangle, it is necessary to analyze what influence each side of the fire triangle has in the combustion process. For the safety manager, this analysis is the key for understanding the concept of fire prevention. Fire prevention attempts to prevent fuels, oxygen, and heat from combining to start a fire. Fire prevention strategies include controlling fuels, controlling oxygen sources, and controlling heat sources. A discussion of fuels, oxygen, and heat sources follows.

Fuel

A fuel is a combustible solid, liquid, or gas. Like in any chemical reaction, a source of energy is needed to sustain the heat required. The most common solid fuels are wood, paper, cloth, coal, etc. Flammable and combustible liquids include gasoline, fuel oil, paint, kerosene, and other similar materials. Propane, acetylene, and natural gas are some examples of gases that are flammable. Solid and liquid fuels share a common characteristic; they must be converted into a gas in order to support combustion. Gaseous fuels can undergo direct oxidation because the molecules are already in the gas state. Some liquid fuels can undergo direct oxidation because they produce vapors at ambient temperatures and pressures. Other liquid fuels and solid fuels, however, undergo sequential oxidation.

This means that a fuel must be heated first to produce sufficient concentrations of gas to support combustion. From a fire safety standpoint, the safety manager should be aware of the different types of fuels located in the workplace.

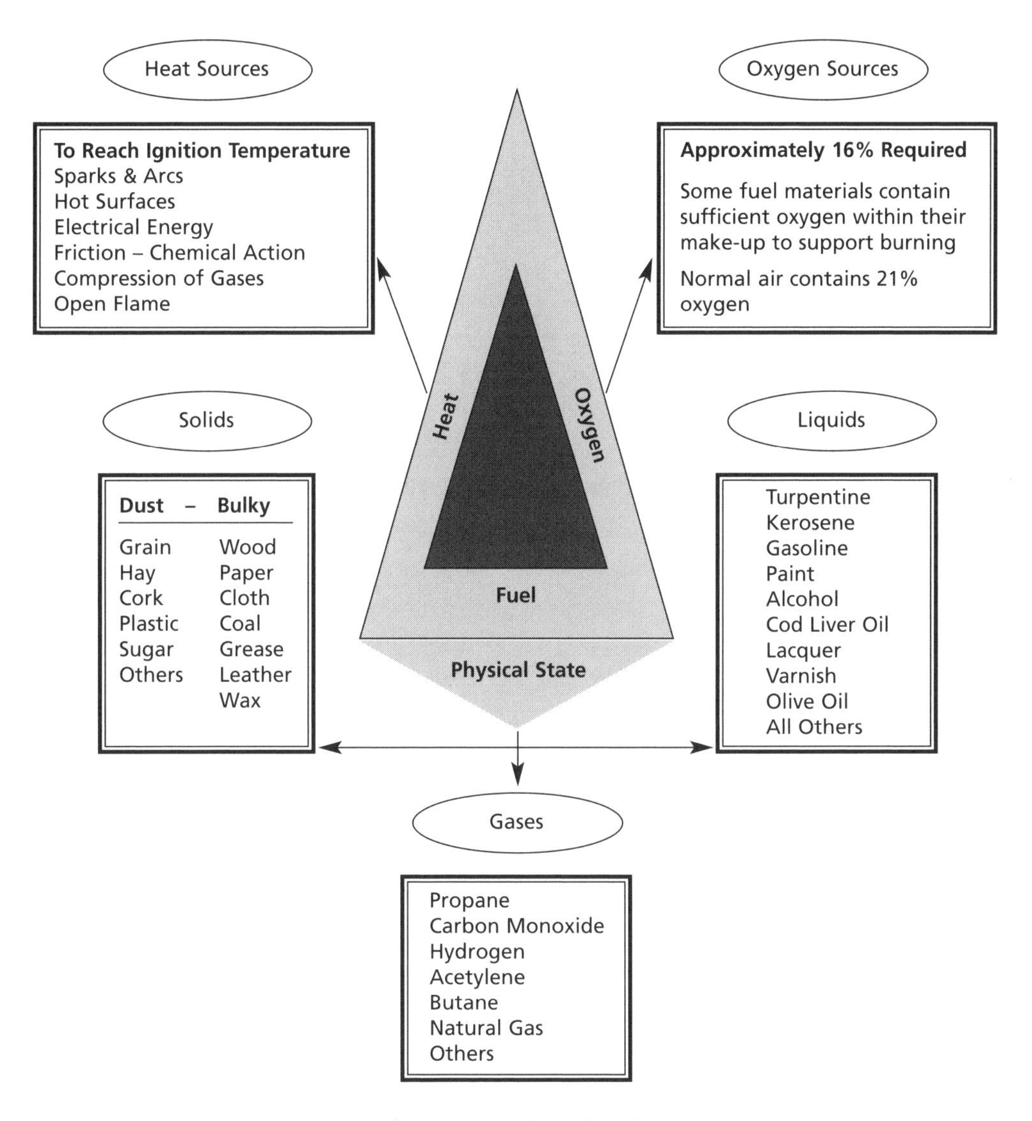

Figure 2–1. Fire Triangle

The ease of ignition of a solid fuel is dependent on several factors. The most important factor is the surface to mass ratio of the fuel. The surface to mass ratio refers to how much of a fuel's surface area is exposed to the environment in relation to its overall mass. The safety manager should be concerned with two things regarding the surface to mass ratio of a fuel. First, the more surface area that is exposed, the easier a fire can initiate and the more rapidly it can burn. Second, the more mass that a solid fuel has the more difficult it will be to initiate and sustain combustion. Consider cotton as a fuel in a textile mill. Cotton dusts and lint will burn easier and faster than a tightly bound bale of cotton.

Liquid fuels are affected by several factors. The safety manager should be familiar with the terms flash point, fire point, boiling point, and specific gravity. Chapter 4 explores these factors in detail. However, one of the most critical indicators of a liquid's flammability should be mentioned—flash point. The flash point refers to the temperature at which adequate vapors are produced to form an ignitable mixture in air. Therefore, a liquid heated to a temperature at or above its flash point will ignite in the presence of an ignition source such as a spark, cigarette, hot surface, or open flame.

Oxygen

The atmosphere contains approximately 21% oxygen by volume. During combustion, the oxygen necessary for oxidation is sufficiently provided from the surrounding air. When the oxygen content of the atmosphere falls below 15%, a free-burning fire will begin to smolder. When the oxygen content of the atmosphere falls below 8%, a smoldering fire will stop burning (Bryan, 1982). Oxygen can also be provided by other sources that release oxygen molecules during a chemical reaction. The safety manager should be aware of these oxidizers in the workplace and segregate them from any fuels.

Heat

The safety manager should be concerned with sources of heat commonly found in the workplace. This is a concern because sources of heat provide the energy necessary to initiate combustion. By preventing heat sources from contacting the ignitable fuel-air mixtures, fires can be effectively prevented from occurring. Some common sources of heat for ignition in the workplace are:

- open flames such as from cutting and welding torches
- cigarettes
- sparks such as from electrical equipment, brazing, or grinding
- hot surfaces such as electrical motors, wires, and process pipes
- radiated heat from boilers or portable heaters
- lightning
- static discharges such as during the transfer of flammable liquids
- arcing from wires and electrical equipment
- compression such as hydraulic oil under pressure on a machine
- exothermic chemical reactions
- spontaneous ignition from slow oxidation or fermentation combined with proper insulation of a fuel

Heat is transferred by three methods: conduction, convection, or radiation. Conduction occurs when two bodies are touching one another and heat is transferred from molecule to molecule. Convection is the transfer of heat through a circulating medium rather than by direct contact. The medium can be either a gas or a liquid. Radiation is the transfer of electromagnetic waves through any medium. For the safety manager, recognizing how heat can be transferred in the workplace is helpful for preventing fires.

FIRE TETRAHEDRON

Fire prevention is the concept of preventing the variables of the fire triangle from coming into contact with each other to initiate a fire. Once a fire begins, it requires four variables to sustain the combustion reaction. The four variables required to sustain a fire are fuel, oxygen, heat, and chemical chain reactions. These four variables represent the fire tetrahedron. The fire tetrahedron is represented in Figure 2–2.

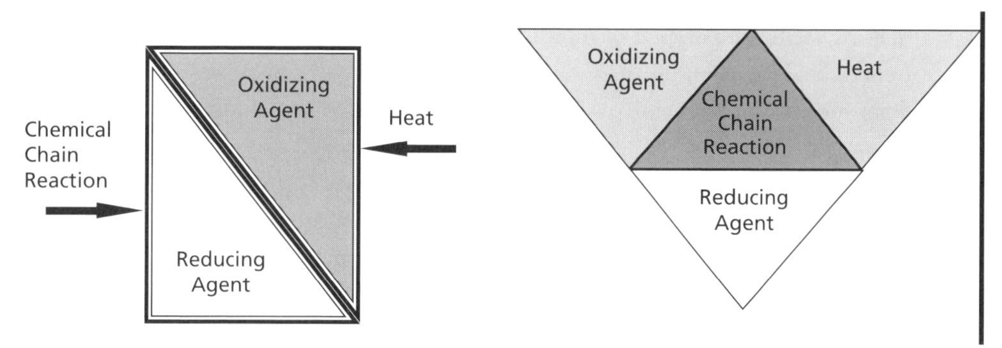

Figure 2–2. Fire Tetrahedron

Chemical chain reactions are a product of the combustion process. The chemical reactions ultimately produce combustion byproducts such as carbon monoxide, carbon dioxide, carbon, and other molecules, depending on the specific fuel. It is these byproducts of combustion found in the smoke that usually affect the safety and health of occupants and fire fighters.

Once a fire begins and is self-sustaining, the goal is to control and extinguish the fire. Fire extinguishment is done by eliminating one of the variables of the fire tetrahedron. By removing the fuel, oxygen, or heat, or inhibiting the chemical chain reactions, a fire can be extinguished. The concept of fire protection assumes fires will occur, and focuses on controlling fires by eliminating or otherwise controlling the variables of the fire tetrahedron. The concept of fire prevention differs from fire protection because fire prevention attempts to control the variables of the fire triangle *before* a fire occurs.

As mentioned, four fire extinguishing principles exist. They are highlighted below (Bryan, 1982).

1. Control the fuel. Controlling the fuel is accomplished by two methods. First, the fuel can be physically removed or separated from the fire. For instance, a fire involving stacks of wood pallets could be controlled by removing any exposed stacks of pallets to a safe location. Another example is closing a valve feeding a gas or

flammable liquid fire. Second, the fuel can be chemically affected by diluting the fuel.

2. Control the oxygen. Controlling the oxygen requires that the oxygen be inhibited, displaced, or the concentration of oxygen be reduced below 15% by volume. Smoldering fires should be diluted to an oxygen concentration below 8% by volume. The oxygen supply to a fire can be inhibited by smothering the fire. Smothering a fire places a barrier between the flame and the atmosphere. This can be accomplished with a blanket or applying a layer of foam to form a vapor barrier. Displacing and reducing the oxygen concentration involves applying an inert gas to the fire, such as carbon dioxide. The carbon dioxide displaces the oxygen thus lowering the concentration to a level that cannot sustain the fire. Applying an inert gas to a fire requires that the fire be located in a confined space. Personnel must be aware that displacing the oxygen or diluting the oxygen concentration affects their ability to breathe. Fire extinguishment using this method requires that personnel be absent from the confined area or protected by self-contained breathing apparatus.
3. Control the heat. Controlling the heat requires that the heat be absorbed. Combustion is an exothermic chemical reaction. If the heat emitted by the reaction can be absorbed faster than the reaction can produce the heat, then the reaction cannot be sustained. Water is the most common extinguishing agent. Water is also the most efficient extinguishing agent because it has the capability to absorb immense amounts of heat.
4. Inhibit the chemical chain reactions. Inhibiting the chemical chain reactions requires that a chemical agent be introduced into the fire. Certain chemical agents can interfere with the sequence of reactions by absorbing free radicals from one sequence that are needed to complete the next sequence. Dry chemical extinguishing agents commonly used in portable fire extinguishers have this ability.

Chapter 7 describes these applications in more detail.

CLASSES OF FIRE

Fires are classified based upon the type of fuel that is consumed. Fires are classified into categories so personnel can quickly choose appropriate extinguishing agents for the expected fire and associated hazards. Fires are classified into four general classes. Each class is based on the type of fuel and the agents used in extinguishment. The four classes of fire are described below:

- **Class A.** Class A fires involve ordinary combustibles such as wood, paper, cloth, rubber, and some plastics. Water is usually the best extinguishing agent because it can penetrate fuels and absorb heat. Dry chemicals used to interrupt the chemical chain reactions are also effective on Class A fires.

- **Class B.** Class B fires involve flammable and combustible liquids and gases such as gasoline, alcohols, and propane. Extinguishing agents that smother the fire or reduce the oxygen concentration available to the burning zone are most effective. Common extinguishing agents include foam, carbon dioxide, and dry chemicals.
- **Class C.** Class C fires involve energized electrical equipment. Nonconductive extinguishing agents are necessary to extinguish Class C fires. Dry chemicals and inert gases are the most effective agents. If it can be done safely, personnel should isolate the power to electrical equipment before attempting to extinguish a fire. Once electrical equipment is de-energized, it is considered a Class A fuel.
- **Class D.** Class D fires involve combustible metals such as magnesium, sodium, titanium, powdered aluminum, potassium, and zirconium. Class D fires require special extinguishing agents that are usually produced for the specific metal.

THREE STAGES OF FIRE

Fires evolve through several stages as the fuel and oxygen available are consumed. Each stage has its own characteristics and hazards that should be understood by safety managers and fire fighting personnel.

Incipient Stage

The incipient stage is the first or beginning stage of a fire. In this stage, combustion has begun. This stage is identified by an ample supply of fuel and oxygen. The products of combustion that are released during this stage normally include: water vapor, carbon dioxide, and carbon monoxide. Temperatures at the seat of the fire may have reached 1000° F, but room temperatures are still close to normal.

Free-Burning Stage

The free-burning stage follows the incipient stage. At this point, the self-sustained chemical reaction is intensifying. Greater amounts of heat are emitted and the fuel and oxygen supply is rapidly consumed. Room temperatures can rise to over 1300° F. In an enclosed compartment, the free-burning stage can become dangerous. Because of the heat intensity, the contents within a compartment are heated. At some point, if the compartment is not well ventilated, compartment contents will reach their ignition temperature. A flashover occurs when the contents within a compartment simultaneously reach their ignition temperature and become involved in flames. It is not uncommon for room temperatures to exceed 2000° F following a flashover. Human survival, even for properly protected fire fighters, is difficult if not impossible for a few seconds within a compartment following a flashover.

Smoldering Stage

The smoldering stage follows the free-burning stage. As a free-burning fire continues to burn, the chemical reaction will eventually consume the available oxygen within the compartment and ultimately convert it into carbon monoxide and carbon dioxide. This causes the oxygen concentration within the compartment to decrease. When the oxygen concentration decreases to 15% by volume, the chemical reaction will not have sufficient oxygen to support free-burning combustion. Visibly, the flames subsist and the fuel begins to glow. A smoldering fire is identified by a sufficient amount of fuels and lower oxygen concentrations. Smoldering fires, especially when insulated within a compartment, can continue the combustion process for hours. Room temperatures can range from 1000–1500° F. The byproducts of combustion also fill the compartment and human survival is impossible. During the smoldering stage, an extreme hazard can develop called a backdraft. A backdraft occurs when oxygen is introduced into a smoldering compartment fire. The immediate availability of sufficient oxygen in the presence of sufficient fuel, heat, and chemical chain reactions causes flaming combustion again. In some cases, the backdraft is so violent that an explosion will occur. Human survival, even of properly protected fire fighters, is usually not possible.

REFERENCES

American Society for Testing and Materials. *ASTM Standards*. Philadelphia, PA: American Society for Testing and Materials (published annually).

Bryan, John L. *Fire Suppression and Extinguishing Systems*. New York: Macmillan Publishing Co., 1982.

Building Officials and Code Administrators, Inc. *BOCA National Building Codes*. Country Club Hills, IL: Building Officials and Code Administrators, Inc., 1995.

Fire Service Extension, West Virginia University. *Firemanship Training*. Section One. Morgantown, WV: Fire Service Extension, West Virginia University.

Friedman, Raymond. *Principles of Fire Protection Chemistry*. 2d ed. Quincy, MA: National Fire Protection Association, 1989.

National Fire Protection Association. *Fire Protection Handbook*. 17th ed. Quincy, MA: National Fire Protection Association, 1994.

______. *National Fire Codes*. Quincy, MA: National Fire Protection Association, 1995.

National Safety Council. Booklets on Fire Prevention and Protection. Chicago, Ill.: National Safety Council, 1982.

Planer, Robert G. *Fire Loss Control*. New York: Marcel Dekker, 1979.

Workplace Safety in Action: Hazard Assessment. Neenah, WI: J. J. Keller and Associates, 1993.

STUDY GUIDE QUESTIONS

True and False

1. True False Class-A fires involve ordinary combustibles, such as wood, paper, cloth, rubber, or certain types of plastics.
2. True False Heat is the energy needed for the fuel to generate sufficient vapors for ignition to occur.
3. True False Fuel is any combustible material—solid, liquid or gas.
4. True False A fire requires four elements for ignition to occur: fuel, oxygen, heat, and the chemical chain reaction.
5. True False Class-B fires include flammable gases and flammable or combustible liquids such as gasoline, kerosene, paint, paint thinners, or propane.
6. True False Fire needs at least 21% oxygen—the same as the air we breathe—for ignition.
7. True False There are four different classes of fires based on the type of objects being burned.
8. True False When fuel and oxygen come together in the right amounts and under the right conditions, a chemical chain reaction happens and fire occurs.
9. True False Class-B fires can be prevented by storing flammable liquids away from spark-producing sources.
10. True False Class-D fires include energized electrical equipment, such as appliances, switches, or power tools.
11. True False Class-C fires include combustible metals, such as magnesium, titanium, potassium, or sodium.
12. True False One way to prevent Class-A fires is to make sure storage and working areas are free of trash.
13. True False Every fire extinguisher displays a rating on the faceplate showing the class of fire it is designed to put out.
14. True False Use the "PASS" method—Pull, Aim, Squeeze, and Sweep—to operate your extinguisher properly.
15. True False If you are trapped in a burning building, never use an elevator.

Other Questions

16. Identify the variables of the fire tetrahedron and explain their relationship.
17 Discuss the three methods of heat transfer and identify an example of each from the workplace.
18. Discuss the three stages of fire.

19. Discuss the three physical states of fuel.
20. The terms fire prevention and fire protection are often used interchangeably. Discuss these concepts and why using the terms interchangeably is incorrect.

CASE STUDIES

1. Locate journal articles for three fires that occurred in the workplace. Analyze each fire and identify the following:
   - the fuels involved
   - the ignition source
   - methods of heat transfer
   - methods of fire extinguishment
   - what fire prevention strategies were employed that failed or were not employed at all
2. Locate and list the various types of fuels, oxidizers, and heat sources in your workplace. Identify the fire prevention strategies that could be implemented to reduce the likelihood of a fire occurring. Also identify the fire protection strategies that could be implemented once a fire does occur.

Essential Elements

CHAPTER 3

CHAPTER CONTENTS

FIGURES

FIRE SAFETY CONCEPTS

A fire safety management program should be based on sound fire safety concepts. Applying fire safety concepts in a logical fashion is the best method for achieving an efficient and cost-effective fire safety management program. The National Fire Protection Association (NFPA) developed *NFPA 550, Fire Safety Concepts Tree*, which can be applied to developing a fire safety system for all structures. Safety managers should use the Fire Safety Concepts Tree for developing and implementing a fire safety management program. Figure 3–1 illustrates an overview of NFPA's Fire Safety Concepts Tree.

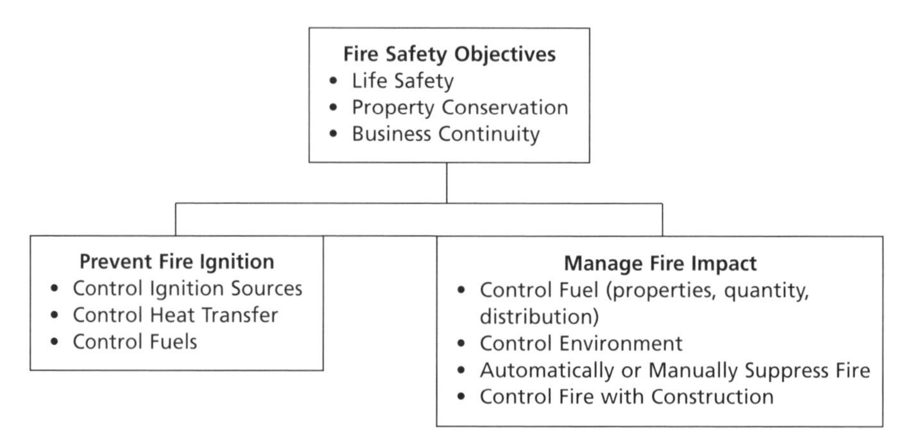

Figure 3–1. Overview of Fire Safety Concepts Tree

This text was written with the Fire Safety Concepts Tree in mind. Individual chapters provide specific information for developing a fire safety management program utilizing the Fire Safety Concepts Tree.

ACTION PLAN FOR DEVELOPING A PROGRAM

A fire safety management program should be developed in an organized fashion. Too often, managers develop programs haphazardly only to wonder why their programs achieve unfavorable results or make little or no impact on an organization. Safety managers should follow good risk management principles when developing any type of program, especially a fire safety management program.

The following sequence should be incorporated into an action plan for developing a fire safety management program:

1. Assess needs and capabilities. This step identifies an organization's fire safety needs and determines what resources are available within an organization and from outside the organization. Needs should be translated into program goals related to life safety, property conservation, and business continuity.

2. Analyze facilities. All of the buildings, structures, and processes located within a facility should be identified and listed. Once listed, the buildings, structures, and processes should be ranked according to the infrastructures' value to the organization. Infrastructure that is crucial to the continued operation of the facility should be given first priority. Next, priority should be given to replacement value.
3. Analyze fire hazards. The ranked list from the previous step can be refined based upon fire hazards within the facility. Target fire hazards such as processes, materials, and environments should be considered. Once a list is developed, the fire hazards should also be ranked based upon the likelihood of a fire occurring and the potential severity of a fire.
4. Develop and implement life safety, fire prevention, and fire protection controls. This step is where the Fire Safety Concepts Tree is truly applied. The result should be a written fire safety management program that includes practices and guidelines for fire risk management through facility design, engineered controls, and administrative controls and activities such as self-inspection, training, education, and communication.
5. Evaluate effectiveness. Once the fire safety management program is implemented, its effectiveness should be evaluated regularly by different organizational levels. Lessons learned should be communicated throughout the organization and improvements made to strengthen weaknesses.

This action plan will help accomplish an organization's fire safety goals in a logical, organized, and cost-effective manner.

PROGRAM GOALS

If an organization is to keep pace with its goals, a proper loss control program requires an effective, well-trained fire brigade, sophisticated detection devices, good housekeeping, and other fire prevention and protection methods included in its programs. These goals can be accomplished by including key elements such as the following.

1. Set policies and establish plans.
2. Create and sustain employee interest.
3. Plan safe buildings, equipment, and processes.
4. Eliminate the causes of fire, explosion, and other losses through proper education, supervision, housekeeping, and maintenance.
5. Provide automatic sprinklers and other protective equipment where needed.
6. Maintain protective equipment in readiness.
7. Organize and train employees for emergency action.
8. Contact the local fire service organization and define their responsibilities for assistance.

Using these goals as guidelines for the fire safety management program, program structure can be designed as shown in Figure 3–2.

PROGRAM ELEMENTS

The complete support of upper management is necessary for the development and implementation of loss control programs. All personnel must participate in a fire safety management program. With innovative methods safety managers can motivate employees to support fire control. A loss control program demands a well-trained fire brigade, effective detection devices, and good housekeeping.

A fire safety management program is divided into eight elements, expressed as performance-based objectives. When objectives are completed, the organization achieves its goals. The eight program elements encompass:

- *Inspections* to detect potential fire hazards, assure that fire protection systems are operable, and assure regulatory compliance. They are conducted by the safety staff or fire brigade members.
- *Education and Training* in the recognition of fire hazards, the use of fire control equipment, and fire code compliance are provided to all employees. Specialized training in fire control equipment and procedures is given to fire brigade members.
- *Fire Suppression* equipment includes extinguishers, sprinkler systems, and alarms. A trained fire brigade can contain and extinguish small fires before they grow large.
- *Emergency Services* include local fire, police, and emergency services. The safety manager develops fire and emergency response plans for the utilization of these agencies. The agencies can serve as consultants for equipment and regulations training.
- *Evaluation of Fire Possibility* is the result of the safety manager's thorough and periodic inspections of layout, inventory, and storage practices.
- *Fire Prevention* is the incorporation of inspections and education to prevent fire losses before they can occur. A comprehensive fire safety management program is an inherent part of a complete organizational structure.
- *Reports and Record Keeping* aid management in fire prevention, suppression, and investigation. Reports describe the needs of the fire safety management program, suggested actions, and actions taken to maintain the program. Records contain inspection and maintenance schedules, fire history, and investigation.
- *Communications* must be maintained between all departments of the organization to insure compliance with the program. The safety manager should establish a good rapport with the local emergency service providers needed during a fire or emergency.

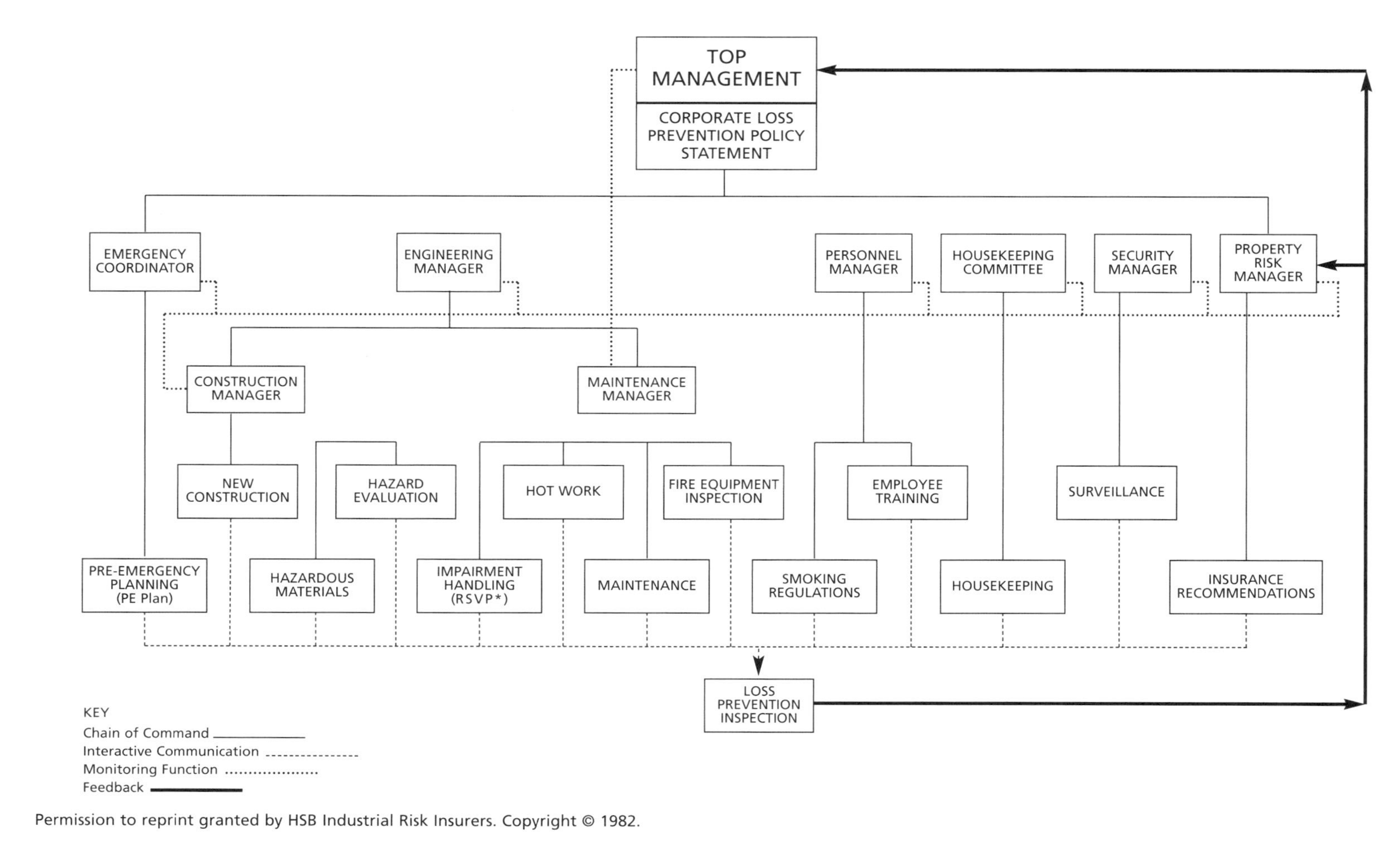

Permission to reprint granted by HSB Industrial Risk Insurers. Copyright © 1982.

Figure 3–2. Loss Prevention and Control Chart

SYSTEM EVALUATION

It is important to remember that since the program has been designed for the organization, and a fire safety management organization has been developed, the initial program step should involve the evaluation of fire possibilities. This evaluation of an organization's facilities will identify fire hazards and establish priorities for the concentration of prevention and control efforts. One evaluation technique that can be effective is the audit survey which can be conducted with assistance from external resources. Support can be bolstered with help from the underwriter's personnel, from insurance sources, or from technical services of brokers. Local fire department personnel can also be used at a lower cost, sometimes free as a public service.

Once the organization has recognized the merit of the evaluation, it is recommended that a form be developed to be used in the audit process. A suggested audit form is shown in Figure 3–3. The audit process should present data related to (1) approximate area that could be involved in a fire; (2) type of construction, combustible or noncombustible; (3) area covered by a sprinkler system; (4) approximate value of building and contents; and (5) relative importance of the building. This information will aid in the identifying and ranking of fire hazards within an organization (Firenze, 1978).

Recommendations

Every phase of the program must be considered in an evaluation. The use of the audit program will emphasize the potential hazards within the organization and show how their programs compare with others. The audit is often difficult, but nonetheless necessary. In this same organization, the safety manager is responsible for formulating and establishing a fire safety management program. The audit must identify the hazards present in the organization (Underdown, 1971). It is a fact that, even though the development of different procedures will rest with personnel having a great deal of fire protection knowledge, the fire safety management program should be coordinated with other personnel within the organization. In order to be comprehensive and effective, all personnel within the organization must have clearly delineated responsibilities to the fire safety management program. Upper management must state policies in writing to support the program and establish responsibility and authority to administer the program. The safety staff of the organization is responsible for staff assistance to line and service departments in fire prevention, fire protection, and control of emergencies affecting the safety of all individuals.

Middle-management personnel are responsible for participating in fire safety management program activities, reviewing fire hazard and regulation compliance conditions, and supporting the establishment of fire brigades. First, line supervision must know the fire protection systems and how they work in their department, cooperate fully in emergency assignments made to their people, and set a good example by working safely. Last, and most important, the line of responsibility lies with the employees themselves. If they are cognizant of their responsibilities to the fire program, many hazards can be reduced and subsequent fires minimized. Employees must comply with supervisors' instructions, report all accidents and injuries, submit recommendations, and know their exact duties in the event of a fire.

XYZ Corporation
Facilities Audit Form

Location: **Date:**

Construction
- Walls: Roof:
- Floors: Partitions:
- Unusual features (combustible interior finishing, insulation, etc.):

Boilers
- Description and rating:
- Fuel:
- Combustion controls (including details on interlocks, purge cycle, flame safeguard, etc.):
- Controls testing program:

Electrical

Power supply (including capacity, no. of feeders, etc.):
- Transformers: No. Capacity
- Major motors: HP Spares?
- Hazardous electrical equipment (Class I or II):
- Emergency generators:
- Data processing: Description
  - Functions
  - Location and envelope construction
  - Tape storage
  - Air conditioning
  - Protection
  - Detectors
- Protection:
  - Computer area
  - Under floor
  - Tape Library

Hazardous operations (flammable liquids, dust, etc.):

Plant protection
- Water supplies (include water test data):
- Underground mains and valves:
- Automatic sprinklers: Full Partial
  - Design
  - Alarms
- Special protective systems (CO_2, Halon, etc.):

Security
- Watchman service: Alarms:

Environs
- Flood: Seismic zone:

Human element program (give date of programs)
- Fire brigade:
- Self-inspections:
- Emergency planning:
- Cutting and welding (hot work):

Reprinted from Planer, p. 28, by courtesy of Marcel Dekker, Inc.

Figure 3–3. Suggested Audit Form

It is essential to obtain backing from upper management for program guidelines. These guidelines will provide the credibility and attention needed to support the fire safety management program (Planer, 1979).

PROGRAM GUIDELINES

After analyzing the information found in the evaluation process of evaluating fire vulnerabilities, protection guidelines can be developed. The fire protection guidelines should include:

1. Incorporation of NFPA codes and standards, OSHA standards, local and state building codes, and requirements for types of building construction necessary for safeguarding life and property.
2. A ranked list of areas which have a significant value to the continuity of the organization's operations.
3. A schedule for conducting life safety evaluations of all the structures within a facility.
4. Design review processes and inclusion of insurance carrier's recommendations.
5. Fire protection assessment schedules including assessment frequencies based upon property values and importance to continued operations.
6. Water supply requirements for high-value areas, or facilities requiring two separate water supplies.
7. Fire protection criteria for high-value areas or facilities requiring a primary means of fire protection to be automatic and secondary means to be manual.
8. Guidelines for protecting special hazards such as computer rooms, hazardous materials, storage warehouses, records archives, and electrical equipment.

Frequently, when large organizations have the need to develop numerous and similar facilities (i.e., warehouses, distribution centers, etc.), it may be advisable to develop corporate loss control guidelines controlling structural building services. The desirability of incorporating guidelines of this nature may be highlighted by an examination of past loss experience which may show a need for improvement.

The safety manager has the responsibility for delegating authority, while the staff will implement the program. This will include assigning various responsibilities to employees, training and education, and evaluating and updating of the program. The line and staff management must assume the leadership of their own department, as well as carrying out their responsibilities. The employees should be fully aware of the policies of the fire safety management program. The delegation of responsibilities will help assure that the program will operate efficiently.

Despite engineering efforts and an established loss prevention program at the corporate level, problems at facilities can continue to exist if the line supervisor does not uphold his/her responsibilities to the fire safety management program. This is recognized by insurance carriers who insist upon implementing management responsibilities' recommendations as a basic requirement for coverage. This assures that activities and efforts such as self-inspection programs, emergency planning, fire brigades, and fire prevention procedures will be implemented. Insurance carriers frequently will insure facilities with major fire protection problems where evidence exists that management is responsive to improvements and is organized. Insurers tend to hesitate insuring facilities where managements are ineffective at implementing and maintaining basic programs.

FIRE BRIGADE

An essential part of every plant emergency organization should be the industrial fire brigade. This organization, as mentioned, may be of an extremely sophisticated nature. A large and valuable facility located in a rural area may require that a sophisticated fire brigade be organized. Conversely, only a small fire brigade that operates portable fire extinguishers may be needed in a facility located in an urban area. A fire brigade should be organized to meet the needs of a specific facility based upon location, the response time of local fire departments, and the value of the facility. *NFPA 600, Industrial Fire Brigades*, is a good resource for organizing a fire brigade. *NFPA 600* has been accepted by the Occupational Safety and Health Administration (OSHA). NFPA currently has a Technical Committee updating the standard for Industrial Fire Brigades' Professional Qualifications. The International Society of Fire Service Instructors (ISFSI) publishes some outstanding performance criteria for industrial fire brigades.

A fire brigade utilizes manual fire fighting methods for fire suppression. This may be considered the entire fire safety effort available to suppress a fire in its early stages. The first objective is to suppress a fire in the event of impairment of automatic protection and to provide extinguishment capabilities where automatic protection is not provided. The point has been demonstrated that a fire brigade will be organized differently for each facility. Manpower requirements for a fire brigade should be established. Anticipated fires should be postulated and the number of persons determined from anticipated tasks required for fire suppression. Management must decide how it will support the fire brigade in all areas, particularly equipment and training.

OSHA requires employers to determine what type of fire brigade they will support. The following scenarios outline the types of fire brigades allowed by OSHA.

1. Employees evacuate buildings only. Portable fire extinguishers are not provided. Employers must provide a written emergency action plan, fire prevention plan, training in evacuation, and shutdown operations.
2. Employees evacuate buildings and portable fire extinguishers are provided for specific, trained employees. Employers must meet the requirements in Scenario 1 in addition to maintaining and testing portable fire extinguishers.
3. All employees can use portable fire extinguishers in their immediate work areas. Employers must provide training in extinguisher selection and use to all employees when first hired and annually thereafter.

4. Portable fire extinguishers will be used by designated employees in assigned areas. Employers must provide a written emergency action plan, fire prevention plan, training as in Scenario 3 and in evacuation and shutdown operations.
5. Portable fire extinguishers will be used by the fire brigade to fight fires in the incipient stage only. Employers must have a fire brigade organizational statement, provide training annually as in Scenario 3 and specific hazards training for brigade leaders and instructors.
6. The fire brigade will fight all fires, including interior structural fires. Employers must have a fire brigade organizational statement and policy, require physical examinations of all brigade members, provide OSHA-required personal protective equipment, provide training annually as in Scenario 3 and in specific hazards for brigade leaders and instructors.

In the selection of a fire brigade leader, it is important that the individual demonstrate good leadership and communication abilities. It was noted earlier that success depends on the ability to fill management positions with people whom other people respect. Assistant leaders should demonstrate similar skills, as they are likely to be in charge of the fire brigade during the absence of the leader. Because it is important that the fire brigade be prepared at all times, it is preferable to use maintenance employees and other nonproduction employees as fire brigade members rather than those committed to production processes.

The fire brigade organizational structure should be broken into squads. Special duties should be assigned to certain squads (such as assuring that control valves are open or fire pumps are running). In addition to extinguishing fires, individual squads have the responsibility for salvage and handling electrical problems. In smaller organizations, multiple functions can be assigned to one squad.

Cragg (1993) outlines the responsibilities and qualifications for fire brigade members and leaders. The *NFPA 600* standard covers the minimum requirements for organizing, operating, training, and equipping industiral fire brigades. Additionally, the standard covers minimum requirements for brigade members.

The fire brigade's responsibilities are to:

- supervise department fire evacuation drills
- operate fire fighting equipment (e.g., extinguishers, hoses)
- provide emergency scene first aid and CPR if needed
- conduct inspections of particular departments
- implement emergency shutdown procedures

Fire brigade members will satisfy the following objectives:

- Explain the function of the fire plan as it relates to his or her department
- Work closely with other members of the fire brigade during fire evacuation drills and actual emergencies
- Demonstrate the ability to extinguish or confine fire with the following:
  a. Class A, B, C extinguishers

b. Carbon Dioxide
c. Hoses

- Demonstrate proficiency in emergency scene first aid
- Demonstrate proficiency in CPR
- List locations of exits, stairways, alarm sirens, escape lights, pull alarms, extinguishers, and hoses
- List evacuation procedures for his/her department
- Be fluent in written and spoken English

The fire brigade leader's responsibilities are to:

- coordinate the implementation of the fire plan
- make certain fire brigade members have received training to satisfy qualifications listed above
- coordinate fire plan into the organization's general safety plan
- assess potential fire dangers to employees
- post building evacuation route maps
- locate emergency first aid centers, if needed
- provide and evaluate emergency scene first aid and CPR
- schedule and evaluate monthly evacuation drills
- schedule monthly inspection of equipment
- schedule monthly inspection of physical equipment (e.g., escape lights, fire doors, illuminated exit signs)
- act as liaison between the organization and local fire and police departments
- delegate a representative in his/her absence

EMERGENCY PLANNING

Emergency planning is essential for avoiding severe losses to people, property, and the continuity of business. A well-developed and rehearsed emergency plan can be the difference for preventing a small emergency from escalating into a catastrophe. Each organization will have certain inherent hazards along with the usual conditions that should be considered during emergency planning. Several resources are available to the safety manager for emergency planning, in particular: *Emergency Management Guide for Business and Industry*, Federal Emergency Management Agency (FEMA); *Hazardous Materials Emergency Planning Guide*, National Response Team; and *NFPA 1620, Pre-Incident Planning*. Insurance carriers and local fire departments are also excellent sources of information for emergency planning.

The first step in emergency planning is to recognize and identify the hazards and determine facility vulnerabilities to emergencies. This can be done by conducting an assessment involving several representatives from different parts of the facility. As discussed previously, the goals are to provide for life safety, conserve property, and assure that business can continue. The potential impact of certain emergencies and the possibilities for long-term interruption of operations should be considered. Emergencies and vulnerabilities should be ranked to prioritize resources later.

The next step is to start planning for the most likely and severe emergencies. The planning process should evaluate the interior layout, escape routes, assembly points, accessibility to fire fighting, ventilation, water supply, detection and alarm systems, communication methods, automatic fire suppression, fire department access, and exposure protection. An emergency plan should detail the duties of personnel and the fire brigade.

The third step in emergency planning is to assure that the organization can implement the emergency plan. Regular fire drills, fire brigade drills, table-top exercises, and full-scale emergency simulations should be conducted. This helps assure that the organization is prepared to respond to emergencies. Rehearsals can also help to identify and correct weaknesses in the emergency plan before an actual emergency. Rehearsals of the emergency plan should also include outside organizations such as fire departments, police departments, hazardous materials teams, industrial rescue teams, and local governments. After an emergency plan is developed and tested several times, it should be critically examined to uncover any further weaknesses that require improvement.

PLANT SELF-INSPECTION

Upper management is responsible for recognizing the loss potential within an organization. Therefore, it should devise a means for identifying those potential losses. One way this may be accomplished is by establishing a plant self-inspection program. A plant self-inspection program is often an important factor for obtaining insurance coverage or reasonable insurance premiums. A self-inspection program should focus on good housekeeping, fire prevention practices, proper maintenance of fire protection features, safe handling of hazardous materials, and other fire safety controls.

Personnel conducting self-inspections should be qualified for the tasks involved. Suggested qualifications might include: a maintenance-oriented background; familiarity with the physical layout of the facility; membership in the fire brigade; and a working knowledge of those items, areas, or processes being inspected. Each organization should determine its own qualifications. Candidates could be enlisted from the safety staff, management, supervisory staff, or hourly employees. They should be the type of individuals who possess the necessary knowledge to perform the task in a reliable and effective manner, as well as personal traits such as a good attitude and responsiveness.

Once personnel have been selected to perform self-inspection tasks, the frequency of self-inspections must be determined. This may vary greatly depending on what is being inspected. The frequencies can vary from hourly, to weekly, to annually. This will depend on the potential fire severity, cost, delay, history of past failures, manufacturer's recommendations, or specific codes and standards. For example, the recommended frequency for inspecting a sprinkler system is weekly, while in an organization with no fixed protection system, monthly inspections are usually sufficient. A sample monthly inspection report is shown in Figure 3–4.

LOSS PREVENTION INSPECTION REPORT

Inspection to be made at least once a month.

Facility: ______________________ Inspector: ______________________

Location: ______________________ Date: ______________________

Identify deficiencies, if any, in the following programs. Make appropriate comments concerning location, specific deficiency, and corrective action taken or required. Major changes in occupancy or construction, as they affect programs, should also be described.

OVERVIEW PROGRAM	Deficiencies None	Deficiencies Noted	COMMENTS
1. Impairments to Protective Systems	☐	☐	______
2. Smoking Regulations	☐	☐	______
3. Maintenance	☐	☐	______
4. Employee Training	☐	☐	______
5. New Construction	☐	☐	______
6. Review of Insurance Recommendations	☐	☐	______
7. Pre-Emergency Planning	☐	☐	______
8. Hazardous Materials Evaluation	☐	☐	______
9. Cutting, Welding, and Other Hot Work	☐	☐	______
10. Fire Protection and Security Surveillance	☐	☐	______
11. Fire Protection Equipment Inspection	☐	☐	______
12. Process Hazard Evaluation	☐	☐	______
13. Proper Housekeeping	☐	☐	______

ADDITIONAL COMMENTS (identify by program number): ______________________

Report reviewed by: ______________________ (signed) Position: ______________________

Figure 3–4. Sample Monthly Inspection Report

Several types of inspections are available for inclusion in a self-inspection program. Each type can be systematic and efficient. The four types of inspection are periodic, intermittent, continuous, and special. Periodic, intermittent, and continuous inspections are particularly useful when incorporated into preventive maintenance programs. Special inspections can be done when new equipment is installed or during promotional campaigns such as the National Fire Prevention Week. Inspection forms are available from many sources in varying detail for different purposes. Insurance carriers, the NFPA, and consulting organizations all have inspection forms available for self-inspection. Many organizations have found that their needs are better met by designing their own forms using other organizations' forms as models.

Upon identifying deficiencies, it is necessary to notify appropriate personnel to correct the deficiencies. Any impairments should also be reported. A system for identifying equipment that is out of service for repair should be developed. Such a system might involve tagging. One part of a tag is left on the equipment needing service. It should be a noticeable color such as red, yellow, or orange. The second part of the tag is sent to the insurance carrier. The third part of the tag is kept by the safety department to indicate that a correction is needed.

CUTTING AND WELDING

Numerous large-loss fires have been caused by portable cutting and welding equipment. Cutting and welding activities can produce sparks, flames, hot slag, or hot pieces that ignite nearby combustibles. Susceptible combustibles are roofing materials, wood, paper, plastics, liquids, and gases. Responsibility for safe cutting and welding rests with the cutter or welder and supervisor (Planer, 1979). Tasks that involve cutting, welding, or brazing are often termed hot work.

Management should establish procedures for approving hot work: designating an individual responsible for authorizing it, requiring the use of approved equipment, and training personnel in hot work procedures. Contractors should be subjected to an organization's hot work procedures also. The supervisor is responsible for assuring that hot work is conducted safely. A typical method for assuring that hot work is conducted safely is through a hot work permit system. Prior to conducting hot work tasks, the cutters or welders must obtain a permit from an authorizing person. Issuance of the permit requires that a pre-task inspection of the work area be conducted by the authorizing person. The pre-task inspection includes observing conditions and housekeeping in the area, protecting combustibles from ignition, and assuring that a fire watch will be posted. A fire watch during hot work is essential. This involves a second person spotting the cutter or welder with a portable fire extinguisher. The idea is that if a small fire is ignited by accident, then the spotter can quickly extinguish the fire before it spreads into a large fire. The hot work area should also be inspected one to two hours following the hot work task to assure that a fire is not smoldering.

NFPA 51B, Fire Prevention in Use of Cutting and Welding Processes, provides further precautions for hot work operations. Figure 3–5 is an example of an inspection form for General Fire Prevention and Protection Requirements for Welding, Cutting, and Brazing.

Side 1

PERMIT FOR CUTTING AND WELDING WITH PORTABLE GAS OR ARC EQUIPMENT

Date ....

Building ....

Dept. .... Floor ....

Work to be done ....

....

Special precautions..........

....

Is fire watch required? ....

The location where this work is to be done has been examined, necessary precautions taken, and permission is granted for this work. (See other side.)

Permit expires..........

Signed ....
(Individual responsible for authorizing welding and cutting)

Time started.......... Completed..........

FINAL CHECK

Work area and all adjacent areas to which sparks and heat might have spread (including floors above and below and on opposite side of wall(s) were inspected 30 minutes after the work was completed and were found firesafe.

Signed ....
(Supervisor or Fire Watcher)

Side 2

ATTENTION

Before approving any cutting and welding permit, the fire safety supervisor or appointee shall inspect the work area and confirm that precautions have been taken to prevent fire in accordance with NFPA 51B.

PRECAUTIONS

- ☐ Sprinklers in service
- ☐ Cutting and welding equipment in good repair

WITHIN 35 FT OF WORK

- ☐ Floors swept clean of combustibles
- ☐ Combustible floors wet down, covered with damp sand, metal, or other shields
- ☐ No combustible material or flammable liquids
- ☐ Combustibles and flammable liquids protected with covers, guards, or metal shields
- ☐ All wall and floor openings covered
- ☐ Covers suspended beneath work to collect sparks

WORK ON WALLS OR CEILINGS

- ☐ Construction noncombustible and without combustible covering
- ☐ Combustibles moved away from opposite side of wall

WORK ON ENCLOSED EQUIPMENT

(Tanks, containers, ducts, dusts collectors, etc.)

- ☐ Equipment cleaned of all combustibles
- ☐ Containers purged of flammable vapors

FIRE WATCH

- ☐ To be provided during and 30 minutes after operation
- ☐ Supplied with extinguisher and small hose
- ☐ Trained in use of equipment and in sounding fire alarm

FINAL CHECK

- ☐ To be made 30 minutes after completion of any operation unless fire watch is provided.

Signed ....
(Supervisor)

Figure 3–5. Example Hot Work Inspection Form

REGULATING SMOKING AREAS

Every year, fire statistics reveal that smoking in the workplace causes severe fire losses. Regulating smoking areas is typically one of management's more difficult employee relations challenges. An employee's desire to smoke is usually in conflict with management's serious concern for fire safety and production efficiency. In most cases, complete prohibition is unrealistic, however, careful regulation can achieve the desired results.

Smoking regulations should be specific. They should clearly state where employees can smoke and when. Smoking should be prohibited in the presence of flammable liquids or gases, combustible dusts, combustible fibers, and substantial storage or processing of combustible materials. The enforcement of no smoking is relatively easy. Although most smokers are careful, a policy on smoking must be established with the careless smoker in mind. The following are some suggested guidelines.

- Focus attention on storage areas. More fires are ignited from careless smoking in storage areas than in any other location. Cartons, packaging, and other combustible materials furnish the fuel needed to initiate a fire. Modern stacking methods favor rapid fire spread thus exposing highly concentrated values.
- Where smoking is permitted in offices, it should be confined to rooms. Smoking in corridors or open spaces should not be allowed. A policy should also be imposed that smoking must cease thirty minutes before the office or room is vacated for the night and smoking materials must be disposed of in proper receptacles.
- Shipping or receiving areas are other locations where careless smoking starts many fires. No smoking should be strictly enforced. Wrapping and packing materials such as carton flats, newspapers, excelsior, burlap, shredded paper, wood bracing, and wooden containers are used in large quantities and provide the fuel for easy ignition.
- All areas should be clearly designated as to whether smoking is prohibited or limited in any way. Where smoking is permitted, such as in generally noncombustible plant areas, cafeterias, rest rooms, and offices, good housekeeping should be emphasized. Ashtrays, sand-filled containers, and similar receptacles should be convenient to smokers. They should be designed to prevent contents from falling out and should not be used for the disposal of general rubbish.

REFERENCES

American National Standards Institute. *Responding to Hazardous Material Incidents*. New York: American National Standards Institute, 1993.

Andersen, Shirley A. *LEAP: A Process for Handling Management Selection and Training*. Technical Communication 37(4) (1990):370–374.

Bird, Frank E., and Germain, George L. *Practical Loss Control Leadership*. Loganville, GA: International Loss Control Institute, Inc., 1985.

Building Officials and Code Administrators, Inc. *BOCA National Building Codes*. Country Club Hills, IL: Building Officials and Code Administrators, Inc., 1995.

Cragg, John R. "Establishing and Managing Fire Services" (unpublished), 1993.

Davis, Larry. "OSHA Standard for Fire Service." *Fire Engineering*, March 1981, pp. 26–31.

Emergency Management Guide for Business & Industry. Washington, D.C.: Federal Emergency Management Agency, 1996.

Establishing and Managing Fire Services Resource Manual. Morgantown, WV: West Virginia University Safety Management Department, 1990.

Firenze, Robert. *The Process of Hazard Control*. Dubuque, IA: Kendall/Hunt Publishing Company, 1978.

Genovese, Robert; Taylor, Trish; and White, Edward. *University of Arizona College of Law Disaster Preparedness Manual*. Buffalo: William S. Hein, 1989.

Hazardous Materials Emergency Planning Guide. Washington, D.C.: The National Response Center, March 1987.

Loss Initiating Adversities Remediation, Morgantown, WV: West Virginia University, n.d.

National Fire Protection Association. *Master Index to National Fire Codes*. Quincy, MA: National Fire Protection Association, 1995.

______. *National Fire Codes, Standard 600, Industrial Fire Brigades*. Quincy, MA: National Fire Protection Association, 1995.

______. *National Fire Codes, Standard 1620, Pre-Incident Planning*. Quincy, MA: National Fire Protection Association, 1993.

______. *National Fire Codes*, vol. 6. Quincy, MA: National Fire Protection Association, 1995.

New York City Transit Authority. *Policy Instruction 8.1.2: Procedures for Response to Rapid Transit Emergencies*. New York: New York City Transit Authority, 1992.

Overview—Management Programs for Loss Prevention and Control. Hartford, CT: Industrial Risk Insurers, 1982.

Planer, Robert. *Fire Loss Control*. New York: Marcel Dekker, 1979.

Tompkins, Neville C. *How to Write a Company Safety Manual*. Boston: Standard Publishing, 1993.

Underdown, G.W. *Practical Fire Precautions*. London: A. Wheaton and Company, 1971.

U.S. Office of the Federal Register Code of Federal Regulations: 29 Parts 1900 to 1910. Washington, D.C.: U.S. Government Printing Office, 1993.

West Virginia University Safety Management Department. *Controlling Environmental and Personnel Hazards Resource Manual*. Morgantown, WV: West Virginia University Safety Management Department, 1979.

Workplace Safety in Action: Hazard Assessment. Neenah, WI: J. J. Keller and Associates, 1993.

STUDY GUIDE QUESTIONS

1. What are the eight program elements that should be included in a fire safety management program?
2. What are the responsibilities of each management level to the fire safety management program?
3. Discuss the formation of emergency plans.
4. What are some specific areas that are inspected where pumps or sprinkler systems are concerned?
5. Who is responsible for safe cutting and welding practices?
6. What specific work areas of an industry should be focused upon when controlling smoking?
7. What are the goals of an effective loss control program?
8. What effect does the organizational structure have upon the fire safety management program?
9. List some of the responsibilities of a fire brigade, fire brigade members, and the brigade chief.
10. List some of the items included in the Protocol for Welding and Cutting.

CASE STUDIES

1. Develop an action plan for implementing a fire safety management program for your employer or university.
2. Develop an emergency plan for the building in which you work or attend classes.
3. Develop a set of program guidelines for a mock refinery, chemical plant, or other high-hazard facility.
4. Form a group of four persons and develop a complete, written, fire safety management program for a facility.

Identification and Control of Materials Considered Hazardous

CHAPTER 4

CHAPTER CONTENTS

FIGURES

IDENTIFICATION OF HAZARDOUS MATERIALS

In the past chemical manufacturers labeled their products with the warnings "Caution," "Danger," "Handle with Care." The terms were vague and did not indicate specific hazards associated with particular chemicals. The U.S. Department of Transportation labeling system (49 CFR 172) contains requirements for the shipping, marking, labeling, and placarding of 1400 hazardous materials. The objectives of this standard are to (1) provide an immediate warning of potential danger; (2) inform emergency responders of the nature of the hazard; (3) state emergency spill or release control procedures; (4) minimize potential injuries from chemical exposure. The standard contains a hazardous materials table listing substances by name, prescribing requirements for shipping papers, package marking, labeling, and transport vehicle placarding.

Table 4–1 shows a comparison listing of United Nations and DOT classifications for hazardous materials. The classes of hazardous materials that must be labeled and placarded are as follows: explosives, flammable and combustible materials, oxidizers, corrosives, poisons, compressed gases, etiologics, and radioactive materials.

The United Nations organization has established a number system to further classify hazardous materials. This class number appears at the bottom of each DOT label. The following table lists these numbers and classifications.

Table 4–1. United Nations and Department of Transportation Classification of Hazardous Materials

United Nations Class	DOT Classification
1	Explosives: Class A, B, and C
2	Nonflammable and flammable gases
3	Flammable liquids
4	Flammable solids, spontaneously combustible substances, and water reactive substances
5	Oxidizing materials and organic peroxides
6	Poisons: Class A, B, and C
7	Radioactive I, II, and III
8	Corrosives
9	Miscellaneous materials which can present a hazard during transport, but are not covered by other classes

Of particular use to safety managers in a fire loss control program is U.S. DOT's *Guidebook for Initial Response to Hazardous Materials Incidents*, published annually. The guide's purpose is to provide instructions for initial action to be taken to protect emergency services personnel (including fire brigade members) in the handling of incidents involving hazardous materials. Table 4–2 from the *Guidebook* presents a discussion of evacuation (isolation) distances involving a chemical spill.

The *Guidebook* cautions its users that it ". . . should not be used to determine compliance with DOT hazardous materials regulations. . . . This guidebook can assist you in making decisions, but you cannot consider it to be a substitute for your own knowledge or judgment. The distinction is important since the recommendations it contains are those most likely to apply to a majority of cases. It is not claimed that the recommendations are necessarily adequate or applicable in all cases. While this document was primarily designed for the use at a hazardous materials incident occurring on a highway or a railroad, it will, with certain limitations, be useful in handling incidents in other modes of transportation and at facilities like terminals and warehouses."

Table 4–2. Table of Evacuation (Isolation) Distances

1. Determine if the accident involves a SMALL or LARGE spill and if DAY or NIGHT. Generally, a SMALL SPILL is one which involves a single, small package (i.e., up to a 208 liter [55 U.S. gallon] drum), a small cylinder, or a small leak from a large package. A LARGE SPILL is one which involves a spill from a large package, or muiltiple spills from many small packages.
2. Determine the initial ISOLATION distance. Direct all persons to move, in a crosswind direction, away from the spill to the distance specified – in meters and feet.
3. Next, determine the initial PROTECTIVE ACTION DISTANCE. For a given dangerous goods, spill size, and whether day or night, try to determine the downwind distance – in kilometers and miles – for which protective actions should be considered. For practical purposes, the Protective Action Zone (i.e., the area in which people are at risk of harmful exposure) is a square, whose length and width are the same as the downward distance.

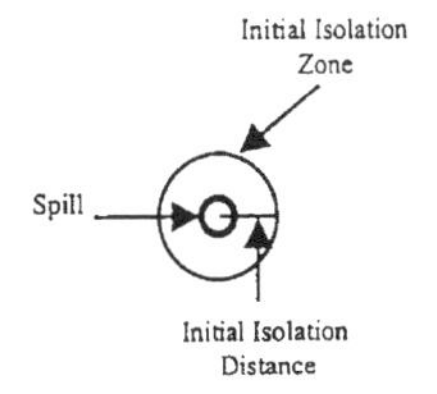

4. Initiate Protectve Actions to the extent possible, beginning with those closest to the spill site and working away from the site in the downwind direction. When a water-reactive PIH producing material is spilled into a river or stream, the source of the toxic gas may move with the currrent or stretch from the spill point downstream for a substantial distance.

The shape of the area in which protective actions should be taken (the Protective Action Zone) is shown in the figure. The spill is located at the center of the small circle. The larger circle represents the INITIAL ISOLATION zone around the spill.

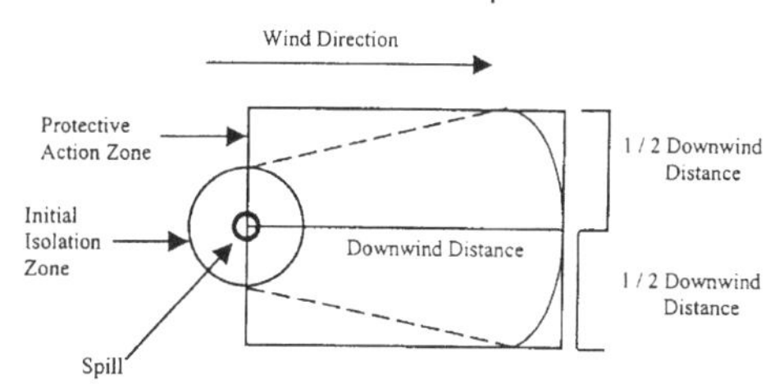

NOTE: See "Introduction to the Table of Initial Isolation and Protective Action Distances" in the *1996 North American Emergency Response Guidebook* for factors which may increase or decrease Protective Action Distances.

Call the emergency response telephone number listed on the shipping paper, or the appropriate response agency as soon as possible for additional information on the material, safety precautions, and mitigation procedures.

Source: *1996 North American Emergency Response Guidebook*.

DOT requires that tank trucks carrying hazardous materials display a numbered placard. The identification number is based on a system created by the United Nations Committee of Experts on the Transportation of Dangerous Goods. The four-digit number identifies a specific hazardous material and has no other meaning (e.g., 1090 represents acetone).

To use the *Guidebook*, identify the material by finding its four-digit number on the placard, shipping papers, or package. The material can be found in the *Guidebook* by number or name. Material name and number refer to a two-digit guide number. The two-digit guide number refers to a detailed instruction sheet in the *Guidebook*. The *Guidebook* also offers instruction to emergency responders in situations where the materials being transported cannot be identified by ID number or name (Figure 4–1).

When approaching a reported or suspected dangerous goods incident involving a placarded vehicle, the following precautionary measures should be considered:

1. **Approach the incident cautiously from upwind to a point from which you can safely identify and/or read the placard or orange panel information.** If wind direction allows, consider approaching the incident from uphill. Use binoculars, if available.
2. **Match the vehicle placard(s) with one of the placards displayed on the following pages.**
3. **Consult the numbered guide associated with the sample placard. Use that information for now.** For example, a FLAMMABLE (Class 3) placard leads to Guide 127. A CORROSIVE (Class 8) placard leads to Guide 153. If multiple placards point to more than one guide, initially use the most conservative guide (i.e., the guide requiring the greatest degree of protective actions).
4. **Remember that the guides associated with the placards provide the more significant risk and/or hazard information.**
5. **When specific information**, such as ID number or shipping name, **becomes available, the more specific guide recommended for that material must be consulted.**
6. **If Guide 111 is being used because only the DANGER/DANGEROUS placard is displayed or the nature of the spilled, leaking, or burning material is not known, as soon as possible, get more specific information concerning the material(s) involved.**

Source: *1996 North American Emergency Response Guidebook.*

Figure 4–1. Recommended Procedure for First Responders to Potential Hazmat Spill

DOT requires color-coded, diamond-shaped, labels on shipping containers holding an amount of hazardous material which could cause a hazardous condition in transportation (see Figures 4–2 and 4–3 on pages 44 and 45). A DOT label indicates the container or package holds a hazardous material. However, because of exemptions based on the size of inner containers placed in outer shipping containers, a package without a DOT label may still contain significant amounts of hazardous materials.

We noted in Chapter 3 that an effective fire loss control program requires an elimination of the causes of fire by education, housekeeping, and maintenance. By identifying potential hazards we can minimize what safety professionals call "hazardous contacts" (i.e., incidents resulting in losses to personnel, property and production). *NFPA 704, Identification of the Fire Hazards of Materials for Emergency Response*, is used for laboratories, chemical processing facilities, warehouses, and storage areas. The system is designed to enable fire fighters to protect themselves from injury when fighting fires in these areas. Obviously, it is necessary for an in-house fire brigade to recognize hazards that exist in their facility.

HAZARD ANALYSIS/CAUSAL INVESTIGATION

This approach is a system-safety effort in the identification and evaluation of all fire loss exposures. Flammable materials are either flammable liquids, solids, or gases. The distinctions are drawn in Table 4–3.

Table 4–3. Classes of Flammable Materials

Hazardous Class	Definition	Examples
Flammable Liquid	Any liquid with a flash point below 37.8° C (100° F).	Gasoline, Pentane
Flammable Solid	Any solid material, other than one classified as an explosive, which is likely to cause fire by self-ignition through friction, absorption of moisture, chemical changes, or retained heat. Can be ignited readily and burn vigorously.	Phosphorus, fish meal
Flammable Solid (Dangerous when wet)	Same definition as above, with the additional fact that water will accelerate the reaction.	Magnesium scrap, Lithium silicon
Flammable Gases	Any mixture or material in a container having an absolute pressure exceeding 40 psi at 70° F or any liquid flammable material having a vapor pressure exceeding 40 psi at 100° F.	Methane, methyl chloride
Combustible Liquid	Any liquid with a flash point at or above 37.8° C (100° F) and below 93.3° C (200° F).	Pine oil, ink, fuel oil

Management's role is to predict, identify, decide, execute, and evaluate the hazards based on sound principles and to be able to understand the nature of the hazards. Since this fire safety handbook's primary purpose is management oriented, a system safety concept to hazard analysis is to identify all physical hazards concerned with the facility. According to Willie Hammer, "a hazard analysis is used to identify any dangers that might be present in a proposed operation, the types and degrees of accidents that might result from the hazards, and the measures that can be taken to avoid or minimize accidents or their consequences." This proactive approach anticipates unsafe actions and conditions.

Hazard analysis/reduction should follow the ten elements of a management program.

1. Identification of Hazards
2. Hazard Inventory
3. Descriptive Information
4. Fire Plan
5. Training Program
6. Inspection Program
7. Scheduled Fire Drills (monthly, annually, etc.)
8. Risk Evaluation
9. Responsibility/Accountability of Management
10. Recommendations at All Levels of Management

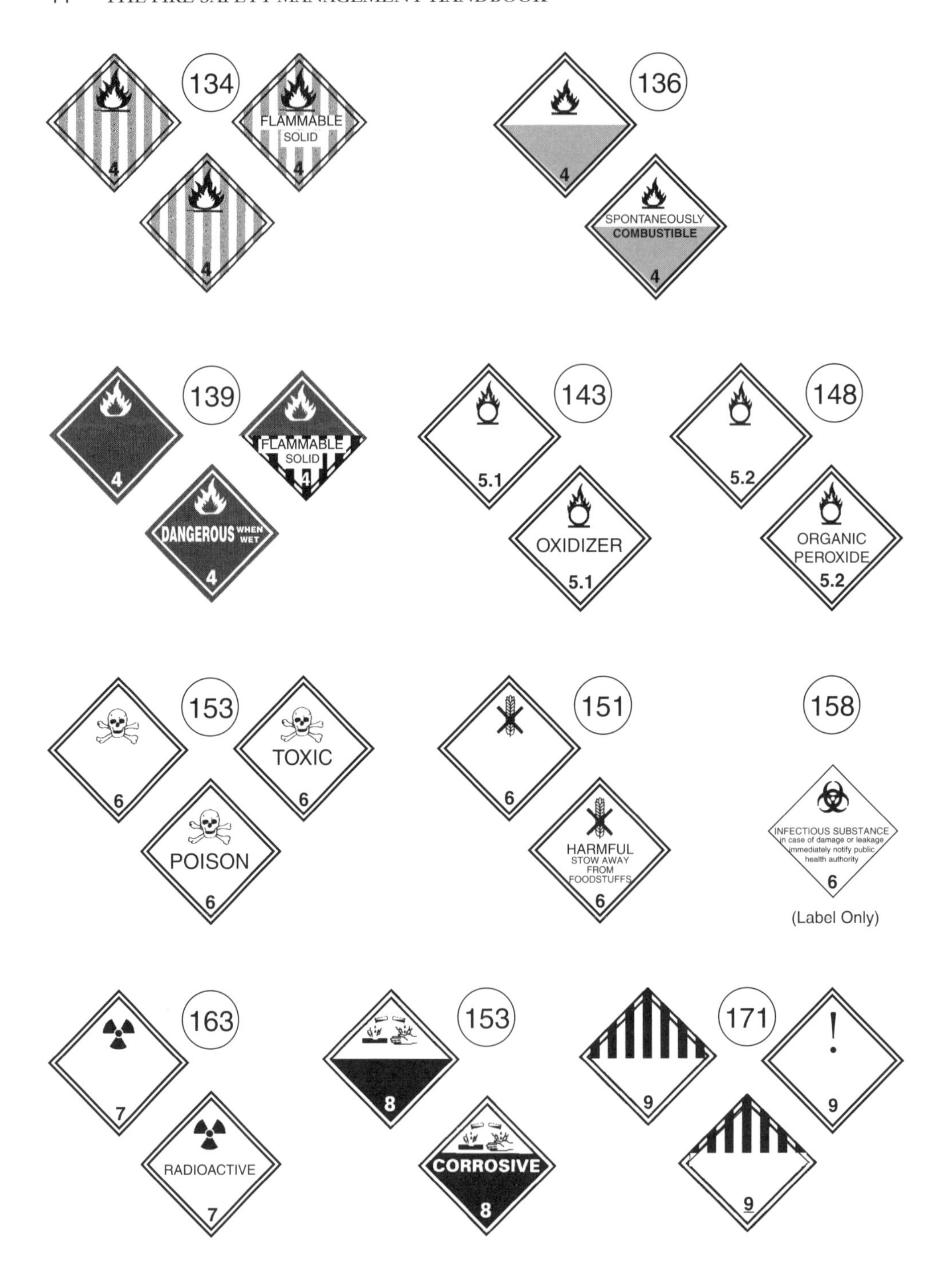

Courtesy, *North American Emergency Response Guidebook*, 1996.

Figure 4–2. Placard Examples

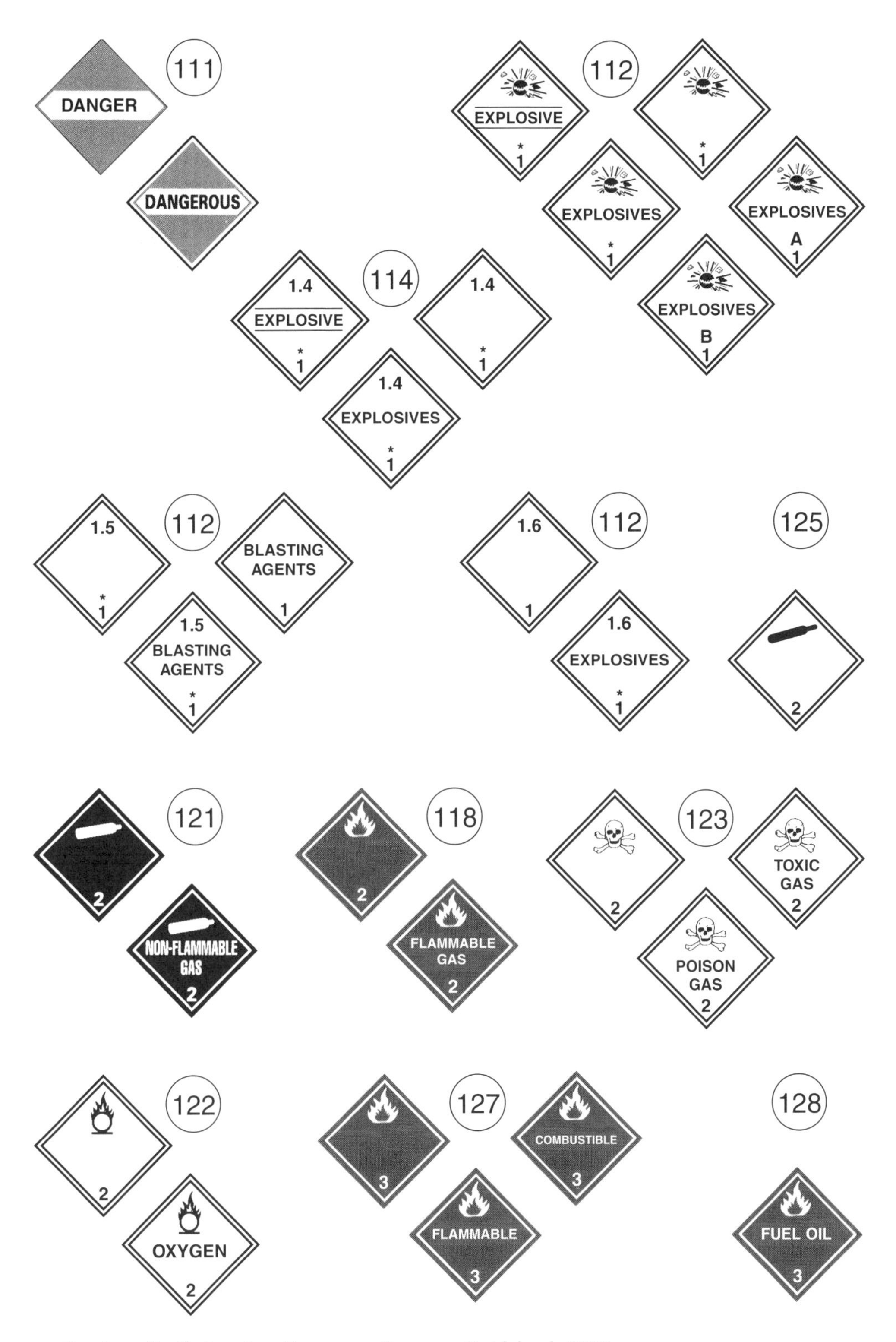

Courtesy, *North American Emergency Response Guidebook*, 1996.

Figure 4–3. Placard Examples

NFPA 704

NFPA 704 provides an easy method of recognizing those hazards. The *NFPA 704 diamond* indicates the health, flammability, and reactivity (i.e., stability) hazards of chemicals by placing numbers in the three upper squares of the diamond (see Figure 4–4).

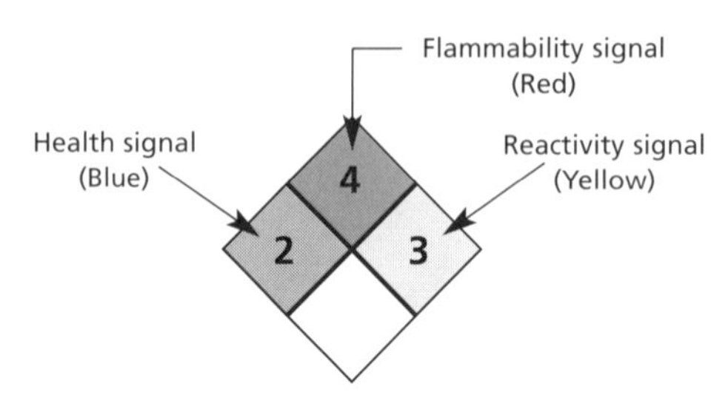

Figure 4–4. NFPA 704 Diamond

Health hazards are indicated in the left square, color-coded blue:

4 Materials which on very short exposure could cause death or major residual injury.

3 Materials which on short exposure could cause serious temporary or residual injury.

2 Materials which on intense or continued, but not chronic, exposure could cause temporary incapacitation or possible residual injury.

1 Materials which on exposure would cause irritation but only minor residual injury.

0 Materials which on exposure under fire conditions would offer no hazard beyond that of ordinary combustible material.

Flammability hazards are indicated in the top square, color-coded red:

4 Materials which will vaporize rapidly or completely at atmospheric pressure and normal ambient temperature, or which are dispersed readily and which will burn readily.

3 Liquids and solids which can be ignited under almost all ambient temperature conditions.

2 Materials which must be heated moderately or exposed to relatively high ambient temperatures before ignition can occur.

1 Materials which must be preheated before ignition can occur.

0 Materials that will not burn.

Reactivity (i.e., stability) hazards are indicated in the right square, color-coded yellow:

4 Materials which in themselves are readily capable of detonations or of explosive decomposition or reaction at normal temperatures and pressures.

3 Materials which in themselves are capable of detonation or explosive decomposition or reaction, but require a strong initiating source, or which must be heated under confinement before initiation, or which react explosively with water.

2 Materials which readily undergo violent chemical change at elevated temperatures, or which react violently with water, or which may form explosive mixtures with water.

1 Materials which in themselves are normally stable, but which can become unstable at elevated temperatures and pressures.

0 Materials which in themselves are normally stable, even under fire exposure conditions, and which are not reactive with water.

Special information is indicated in the bottom square, color-coded white:

1. The letter W with a bar through it indicates a material may have a hazardous reaction with water. This does not mean "use no water," but rather "avoid the use of water." Note that some forms of water (e.g., fog or fine spray) may be used. Because water may cause a hazard, it is advised that water be used very cautiously until fire fighters have proper information.
2. The radioactive "pinwheel" indicates radioactive materials.
3. The letters "OX" indicate an oxidizer.

MSDS

The OSHA Hazard Communication Standard (29 CFR 1910.1200) requires Material Safety Data Sheets (MSDS) in addition to labels and warnings on containers. The standard assigns responsibility to the chemical's supplier for the content of the MSDS. Employers are required to do the following:

1. Make MSDSs available to all employees, designated representatives, and OSHA.
2. Instruct employees to label all portable containers containing hazardous substances not intended for their immediate use.
3. Train employees about the main features of the Hazard Communication Standard: labels, MSDSs, the company's written program, list of hazardous materials, and required training. The employer is required to instruct employees in how to recognize, understand, and protect themselves from hazards they will encounter in the workplace.

It is the employer's responsibility to confirm that the MSDSs supplied with chemicals are adequate. If not, the supplier must be informed and any inadequate or incomplete information corrected. The information must be written in English. The following minimum information is required on a Material Safety Data Sheet:

1. Specific identity of each hazardous chemical or mixture ingredient and common names
2. Physical and chemical characteristics of the hazardous materials such as:
   a. Density or specific gravity of liquid or solid
   b. Density of gas or vapor relative to air
   c. Boiling point
   d. Melting point
   e. Flash point
   f. Flammability range
   g. Vapor pressure
3. Physical-hazard data such as stability, reactivity, flammability, corrosivity, explosivity
4. Health-hazard data including acute and chronic health effects and target organ effects
5. Exposure limits such as OSHA Permissible Exposure Limits (PELs)
6. Carcinogeneity of material
7. Precautions to be taken including use of personal protective equipment
8. Emergency and first aid procedures, including spill cleanup information and the EPA's reporting requirements in the event of a spill
9. Supplier or manufacturer data including:
   a. Name b. Address c. Telephone number d. Date

No blanks are allowed in any section of an MSDS. Note that the OSHA Hazard Communication Standard is performance oriented, that it is the responsibility of the employer to develop the specifics of a HazCom program. Because of this there is no standard format for Material Safety Data Sheets. Mansdorf (1993) notes that "...for 20 different suppliers you will find 19 varied formats for MSDSs. The layout, orientation, type fonts, order of information, categories, comprehensiveness, length and almost any other attribute are widely variable. Some MSDSs are encyclopedic in treatment and may run to 20 pages while others will be only one or two pages with very brief information."

Keep in mind that information must be imparted from MSDSs to each employee so that he/she learns at least the following about each chemical in his/her work environment:

1. How hazardous is it? (a) Flammable? (b) Toxic? (c) Poisonous? (d) Explosive?
2. Will it cause cancer?
3. What organs does it affect (e.g., lungs, skin, etc.)?

4. Will it evaporate readily?
5. Will the liquid float or sink in water?
6. Will the gas vapor rise or fall in air?
7. What are the exposure limits?
8. What protective equipment must be used?
9. What first aid measures are required if exposed?
10. What to do if there is a spill.
11. How to handle a spill.
12. Whom to phone or write to for more information.

Making available an MSDS does not satisfy the need for necessary training. One loose-leaf notebook of MSDSs is inadequate where multiple chemicals are used at multiple work stations. It is wise to keep alphabetical files of the most recently available MSDSs for particular chemicals used in each department. A permanent indexed file should contain all MSDSs for all materials. Only MSDSs for materials used in a particular department should be made available to the employees of that department.

EMERGENCY RESPONSE MUTUAL AID NETWORK

Because many jurisdictions and in-house fire brigades are not equipped to deal with hazardous materials incidents, the Chemical Manufacturers' Association established the Chemical Transportation Emergency Center (CHEMTREC). CHEMTREC was created in 1971 in response to a growing need for information and assistance by emergency responders and carriers of hazardous materials. The center was developed as both a resource for obtaining immediate emergency response information and as a means by which first responders (i.e., local fire departments and in-house fire brigades) could obtain prompt technical assistance from chemical industry experts during emergencies.

CHEMTREC is in operation 24 hours a day, seven days a week, and can be reached through its toll-free emergency number, (800) 424-9300, from anywhere in the United States, Puerto Rico, the U.S. Virgin Islands, and Canada. CHEMTREC can be reached from outside the United States and Canada and from ships at sea by calling collect the international and maritime number, (202) 483-7616. The emergency number is displayed on tank trucks and rail cars to assist first responders on the emergency scene.

In addition to CHEMTREC, the Chemical Manufacturers Association operates an emergency response mutual aid network made up of emergency response teams of member companies and commercial contractors. The network is activated when CHEMTREC notifies a member company that an incident involving one of its products has occurred and expert assistance may be needed at the scene. If a shipper that is a member of the network is unable to respond in a reasonable time because of distance or other circumstances, CHEMTREC links the shipper with a response team closest to the scene that can provide assistance. The team, which could be another chemical company's or that of a private contractor, responds on behalf of the shipper and provides assistance until the shipper's personnel arrive.

CHEMTREC has a reference library of 1 million MSDSs and has several mutual aid programs for specific products such as chlorine, phosphorus, hydrogen fluoride, hydrogen cyanide, vinyl chloride, sulfur dioxide, hydrogen peroxide, compressed gases, and swimming pool chemicals. CHEMTREC's public service non-emergency number provides health, safety, and environmental information. Call toll free from anywhere in the United States, Monday through Friday, 9:00 AM to 6:00 PM eastern time, (800)262-8200.

Cote and Bugbee (1988) caution that ". . . releases of hazardous substances create incidents with a wide variation of consequences, ranging from incidents with little if any impact to those affecting thousands of people and with the potential for severe environmental effects." Because of this, ". . . plans for responding to hazardous material incidents should be prepared in advance. To respond safely and effectively, appropriate equipment should be available, personnel should be trained, and necessary resources should be readily available. Advance plans should be reviewed, tested and kept current."

ISOLATION OF HAZARDS

When involved in the planning stages of an industry, it is important to plan for the control or protection of hazardous materials. These hazardous materials should be isolated from other areas such as power supplies, storage areas, or emergency exits of the plant. The planner must determine if the hazardous material, regardless if it is to be stored or used for processing, can be located in an open area away from the work site, or if it can be placed in a detached structure (Meidl, 1978). Cost is an important factor in any decision made. Purchasing land for additional storage or processing could be quite costly for an enterprise. Additional structural costs for storage or processing will also be a factor. For the more volatile and dangerous operations, however, the need for greater fire safety will frequently be an offsetting factor to economic considerations. Also, the laws and codes of different cities and states may require certain specifications.

Where space is limited, enclosures that physically separate the hazard from other areas should be provided. The enclosure must be of a sufficient fire resistance rating to separate or confine the fire in case of ignition. In order to determine the fire resistance needed for walls or other structures, it is necessary to know the amount of combustible or flammable material within an area. This is known as fire loading. In Class A materials, measurements are determined by obtaining the total weight and dividing by the area to obtain an average in pounds per square foot (psf). Most Class A materials have heats of combustion from 7000 to 9000 BTU/lb., while flammable liquids have approximately twice that amount. The British have conducted considerable research over the years on fire loading. They have developed three classifications: low, moderate, and high. The design of structures can be related to the caloric content of materials stored within. By using this system, the severity of the fire can be roughly determined. The fire resistance ratings as listed by Planer (1979) are:

Low Fire Loading: Not exceeding 100,000 BTU/ft^2, 1-hour fire resistance. This classification will generally apply to schools, offices, hotels, and most areas of hospitals.

Moderate Fire Loading: Between 100,000 and not exceeding 200,000 BTU/ft^2, 2-hour fire resistance. This category will encompass most industrial occupancies such as machine shops and assembly areas. The assembly area is where a collection of parts are assembled into a complete product, ready for shipping.

High Fire Loading: Exceeding 200,000 BTU/ft^2, but not in excess of 400,000 BTU/ft^2, 3-hour fire resistance. Generally, this classification will cover warehousing of various products.

There is also a special class for those occupancies having loading in excess of 400,000 BTU/ft^2. Flammable liquids will most likely fit into this particular classification.

As was stated earlier, the economics of fire-resistive walls can be quite prohibitive. As a consequence, more reliance is placed upon protective systems, like complete automatic extinguishment systems. When addressing the subject of enclosures to confine flammable liquid hazards, consideration must be given to openings in fire walls or fire barriers which could allow the substance to escape from the enclosure to the surrounding area. Several measures can be taken to prevent the escape. First of all, regular maintenance can be conducted on the area to repair or replace any holes, cracks, etc., that may appear. Drainage trenches across all openings can also be considered. The safe handling of flammable liquids also needs to be considered. Liquids must be confined whenever possible. There are various sizes and shapes of safety containers, and these are used. The use of "special devices" when loading or unloading tank cars so that a "closed system" is maintained is also recommended (Factory Mutual Engineering Corp., 1989).

It is, of course, desirable to isolate hazards and confine flammable liquid spills to the hazard area. The safest method of handling flammable liquids is by enclosing piping systems properly so that in case of physical damage to the piping which causes leakage, the liquid will not continue to flow either by gravity, by a pressurizing agent, or siphoning. Pumps properly designed provide a safe shutoff to prevent siphoning when not in operation. Table 4–4 provides the NFPA's classification of flammable and combustible liquids.

Table 4–4. Classification of Flammable and Combustible Liquids (*NFPA-30*)

Class I	**Flammable Liquids** - Flash point below 100° F (37.8° C)
	Volatile Class I Flammable Liquids
	Class IA - Most hazardous, having flash points below 73° F (22.8° C) with boiling points below 100° F (37.8° C)
	Class IB - Same flash point range but with boiling points at or above 100° F (37.8° C)
	Class IC - Flash points between 73° F (22.8° C) and below 100° F (37.8° C)
Class II	**Combustible Liquids** - Flash points at or above 100° F (37.8° C) and below 140° F (60° C)
Class III	Liquids are included in the combustible liquid classification and are further classified:
	Class IIIA - Flash point between 140 and 200°F (60–93.4°C)
	Class IIIB - Flash point 200° F (93.4° C) or above

Emergency shutdown systems also have to be considered. These should include designing for both human and mechanical failure. When using positive displacement pumps, a relief valve of adequate capacity must be used to prevent over-pressurizing the piping. When pumps are handling flammable liquids, they must be located outdoors away from main buildings unless designed for indoor use. If this isn't possible, they could also be located in a separate pump house or cutoff room. These areas must be constructed of noncombustible materials.

The material used for the actual piping must be resistant to the corrosive properties of the liquid handled. Also gaskets, flanges, and joint compounds should be selected with care to avoid weak points in the piping and pump system.

In addition, the location of the piping system is important. It should not be located in main building areas. Pipes can be run outside either underground or along the side of the building. If the piping system is installed inside a building, vital areas must be covered with sprinkler systems and cutoff valves should be located in an easy access area away from the vital areas.

When storing large amounts of flammable liquids, special care is needed because of the potential of large losses. The tanks should be located away from main areas, should be easily accessible and should be properly ventilated in accordance with nationally recognized standards.

Storage tanks are usually constructed of steel. Other materials can be used depending on local codes or if the liquid is a corrosive element to the steel. The tanks should be of noncombustible construction when used above ground and conform to nationally recognized standards.

Tanks are classified according to pressure, either atmospheric, low-, or high-pressure vessels. The design of horizontal tank supports should be such that it will not tilt or collapse under fire conditions. Firm foundations are necessary.

Outside storage tanks should be located away from property lines or other important buildings. *NFPA 30* lists these minimum distances. In some cases, however, additional space should be required.

In above-ground tank farms, the tanks should be located on ground that drains away from important buildings. If possible, a slope of not less than 1% away from the tank toward the drainage system should be provided. In lieu of drainage trenches for protection, a dike could be constructed around the tank. Dikes can be made of earth, steel, concrete, or any material that will resist the penetrating or corrosive nature of the liquid. Heights of the dike should be restricted to an average height of six feet above the interior grade to prevent entrapment of personnel.

All tanks should allow for ventilation as well as equalization of the pressure changes created by variations in temperature. How much venting is needed depends upon filling and emptying pumping rates. Specific requirements can be found in the *American Petroleum Institute's Venting Atmospheric and Low-Pressure Storage Tanks* (Standard 2000).

The more volatile flammable liquids require venting devices which are normally closed except when tanks are under vacuum and pressure. In Class I liquids, the venting devices should be provided with flame arresters except when condensation, crystallization, or freezing make the use of flame arresters impractical. The purpose of the flame arresters is to prevent flashbacks of fire to tanks where the vapor mixture may be in the explosive range.

There should also be provisions for emergency relief venting. The tank can be constructed with a weak seam on the top to allow for venting. For calculation of adequate emergency relief venting, check *NFPA 30*.

Perhaps the safest way of handling flammable liquids is the outside storage tank. Good judgment must be used in the location of these tanks.

Because they are difficult to ventilate, tanks should not be located inside buildings. If there were ever an explosion or fire in one, it could destroy the building and perhaps cause other tanks to catch fire.

As was stated earlier, the safest way of handling flammable liquids is by an enclosed piping system. However, this is not always possible or probable. When liquids are stored in drums of other types of portable tanks, preferably they could be in a separate building at least 50 feet away.

When storing containers of over 30-gallon capacity, *NFPA 30* recommendations should be evaluated. When dispensing the liquid from a drum or portable tank, approved pumps or a self-closing faucet should be used.

When storing several drums or portable tanks indoors, the need for explosion-relief building design should be considered. This prevents a second explosion if one section or drum explodes. It should also be kept in mind that tanks or vessels used for processing should be provided with fire-resistive or protected steel supports. Otherwise an exposure fire of sufficient duration could cause the collapse of the tanks or vessels.

IGNITION SOURCES

Ignition is a means of either an intentional or unintentional source that provides for the initiation of self-sustained combustion. Limitations must be set on all types of ignition. There are many different ignition sources. Some sources of ignition include heated surfaces, open flames, smoking, frictional heat, static, radiant heat and cutting and welding that is a heat-producing action (*NFPA Fire Codes*, 1995). There are basically four categories consisting of electrical, chemical, mechanical, and nuclear source of ignition. Electrical ignition sources are based on resistance, static electricity, arcing, sparking, and lightning. The chemical process of ignition can be categorized into combustion, spontaneous heating solution, and decomposition. Generally, chemical ignition is when something is burning, this is called combustion and it is generated by heat. Mechanical ignition sources occur in two ways: friction and compression. The friction process is generated when the action of two surfaces in motion creates a heat buildup and eventually the heat produces sparks that ignite combustibles that are close by. The last of the four categories, nuclear, occurs when the heat is generated by fission (atoms are split apart) or fusion (joined) of more than one nuclei. During nuclear explosions the high pressure is created by the large quantities of heat produced by fission and fusion.

FLASH POINT AND IGNITION POINTS

Flash point is the temperature at which a flammable liquid produces enough vapors to be ignited in the presence of air. This is the most significant property of all. The flash point determines the flammability of a liquid more than any other single factor. The reason flash point is so important is that flammable liquids are not ordinarily flammable; it is the vapors they produce that ignite or explode. NFPA defines flash point as "the lowest temperature of the liquid at which it gives off vapor sufficient to form an ignitable mixture with the air near the surface of the liquid or within the vessel used." The liquid will not support continuous combustion until its temperature has reached the fire point. The fire point is usually a few degrees higher than the flash point. Since some liquids can generate explosive vapors at the flash point, both points are important to the firefighter.

In summary, if the temperature is high enough, every flammable liquid will produce vapors within the flammable range. If an ignition source of sufficient intensity is present, the vapors will ignite or explode. Methods used to combat the hazards of flammable liquids must be based upon the answers to the following questions:

1. What is the flash point of the liquid?
2. Is the liquid or vapor soluble in water?
3. What is this liquid's specific gravity?
4. What is the vapor density of this vapor or gas?
5. Where are the ignition sources located?

Liquids with flash points at or below ambient temperature are easy to ignite, and burn rapidly. Liquids with flash points above ambient temperatures have less of a risk based on the fact that they must be heated in order to ignite. Flash point determinations give rise to hazard classifications systems, the most severe hazard being afforded by those liquids with the lowest flash points.

Ignition Temperatures

Ignition is the process by which self-sustaining combustion is initiated. All flammable liquids, solids, and gases have an ignition temperature. NFPA defines this as "the minimum temperature required to initiate or cause self-sustained combustion independently of the heating or heating element." (In other words, anything flammable will burn at a certain temperature without an additional ignition source.) This differs from the flash point in that the flash point only tells us when sufficient vapors will be present to burn. They still require a source of ignition.

Considering first a flammable gas or vapor-air mixture, piloted ignition can be achieved by the introduction of an ignition source, such as a flame or spark. However, if the temperature of the mixture is raised sufficiently, the mixture will undergo autoignition, in which the onset of combustion is spontaneous.

STATIC ELECTRICITY

In some cases grounding may help prevent surface charges. All equipment containing Class I flammable liquids should be electrically grounded. This is particularly necessary at points where the liquid is transferred from one container to another.

There are other ways to dissipate static charges. One method used in hospital operating rooms is humidification. This is by no means foolproof.

Ionization is another method used. This method is used on nonconducting objects such as rubber or leather conveyor belts, or on paper or cloth in coating operations.

GASES

Flammable gases have many of the properties, as well as problems, of flammable liquids. Fire fighters know that burning gases under pressure must not, under any circumstance, be extinguished unless it is possible to control the gas flow.

Many types of gases are shipped in a liquefied form under low temperatures and higher pressure. With higher temperature or a release of pressure, the liquefied gas will return to its gaseous state. Also, gases stored in this sealed, liquefied state will occupy a smaller volume of space. In the instance of liquefied natural gas, its volume would be 600 times greater in its natural form.

Vapor Density

This can be defined as the relative weight of the vapors of a liquid as compared with air. This property is an important factor to consider when designing a ventilation system to remove hazardous vapors from a work space. All flammable liquids have vapor densities that are greater than one (1). This means that the vapors will settle low and may travel along floors and accumulate in pits and trenches. Therefore, to be effective, the ventilation systems must be designed to remove heavier-than-air vapors.

Liquefied Petroleum Gas (LPG)

Liquified petroleum gases are derived from oil refinery gas. It is composed of certain specific hydrocarbons which can be liquefied under moderate pressure at normal temperatures but which are gaseous under normal atmospheric conditions (Isman and Carlson, 1980).

LPG is liquid under pressure, but vaporizes upon the release of pressure. LPG is stored and transported as liquid and is stored and handled with the use of lower temperatures and higher pressures. LPG is stored in pressure vessels. LPG includes Propane, Butane, Propylene, and Butylene.

Propane has a very high vapor pressure which at 70° F is 124 psi and at 100° F is 192 psi. It can be stored under pressure without refrigeration or with only moderate refrigeration. Propane liquefies at –44° F and supplies 2% of the nation's energy. Propane is a flammable, compressed gas that is also the most commonly used LPG in industry today. Class B- and C-rated extinguishers can be used to control or extinguish this type of fire.

Butane has a lower vapor pressure than propane. It has a vapor pressure at 70° F which is 31 psi and at 100° F is 59 psi. Butane liquefies at –31° F and is the second most commonly used LPG today. Class B- and C-rated extinguishers can also be used to control or extinguish this type of fire.

LPG is pressure-stored at 10° F and warmer. A single-wall tank is required for use when storing LPG at this particular temperature. Temperatures colder than 10° F may require a double-wall tank. Storage containers for LPG have a capacity in the range of less than 125 gallons to 90,000 gallons and should meet design requirements of the American Society of Mechanical Engineers (ASME). Eighty-five percent of LPG is stored under pressure in underground salt domes of mined caverns. LPG can also be stored in bottle-type gas system containers, construction of which is approved by the Department of Transportation (DOT).

Regulators or pressure-reducing valves are used to control distribution by utilizing pressure. They should be located as close to a container as possible and should be installed inside buildings.

The excess flow valve has a pressure differential that is produced by an increased flow through the valves. This valve is a spring-loaded type, with a small bleed hole to permit pressure equalization. Excess flow valves are designed to close at 150–200% of a normal flow. They must be installed in the correct position to close with the direction of flow and should be tested at five-year frequencies for proper closing action.

Back pressure check valves should be provided on fill connections. They are a spring-loaded-type valve or weight-loaded-type valve with an in-line swing action which is designed to close when flow is either stopped or reversed.

Some safety relief valves are also spring-loaded types which are subjected to clogging. To prevent this, a safety relief valve with a three-way valve is installed beneath two safety relief valves, permitting one valve to be open at all times. Each safety relief valve must be plainly and permanently marked as required by *NFPA 58, Liquefied Petroleum Gas Code* (*National Fire Codes*, 2001).

Department of Transportation (DOT) cylinders are gas-type, bottled containers that have a fusible plug. These cylinders are also spring-loaded types.

Vaporizers are needed for LPG having a low vapor pressure such as butane. They should be located outside or in noncombustible vaporizer houses. Indirect-heated types of vaporizers should be located a minimum of 20 feet from openings in walls of buildings.

Direct-fired vaporizers should be located a minimum of 50 feet from openings in walls of buildings. They also should be a minimum of 75 feet from storage container relief valves. The National Fire Protection Association requires that direct-fire-type vaporizers be a minimum of 10 feet from a container, 15 feet from container shutoff valves, 20 feet from point of transfer, and 25 feet from the nearest building.

Acetylene

Acetylene is used extensively as a raw material in the chemical industry. It is also used in welding and cutting. Acetylene is colorless and odorless and has an ignition temperature of 571° F. The explosive range of acetylene is 2.5–100% in a given volume of air (Planer, 1979).

Acetylene generators are sometimes found in industry. Generally, these generators are of *carbide-to-water* or *water-to-carbide* types. For safety reasons carbide-to-water-type generators are generally used. In the event of an acetylene fire resulting from wet calcium carbide, extinguishment is recommended by using dry powder or carbon dioxide extinguishers. No water or water-based foams are to be used in such situations.

The sensitivity of acetylene to explosion is resolved in cylinders by dissolving compressed acetylene with a special porous filler. Acetylene cylinders, when used, must be properly marked and shall be legibly marked with the name of the gas in accordance with ANSI 248.1 (*Standard Method of Marking Portable Compressed Gas Containers, NFPA 51-2-13*). They should be stored and secured in an upright position. Cylinder valves should be closed on empty cylinders while in storage or shipment. Acetylene cylinders should be kept at least 20 feet from combustible materials, and caps to protect valves should be used while not in service. Defective cylinders should be removed from service. Warning signs should be placed in areas where acetylene is used, and proper fire fighting equipment should always be conspicuous and readily available (NIOSH, 1976).

Oxygen

Oxygen is colorless and odorless. Although nonflammable, oxygen is a supporter of combustion to such an extent that certain materials, such as oils and other combustible lubricants, may ignite explosively at ordinary temperatures in the presence of pure oxygen. Materials thought to be only slightly combustible will burn intensively in oxygen. Use of oxygen and other fuel gases such as acetylene, natural gas, liquefied petroleum gas, hydrogen, and methylacetylene-propadiene (MAPP) is common during welding and cutting operations. To a lesser degree they are used in heat treating and other processing

The National Fire Protection Association has developed what is known as *Standard 51, Oxygen-Fuel Gas Systems for Welding and Cutting*. OSHA also requires compliance to the mandatory section of the standard.

NFPA Standard 51 deals exclusively with oxygen-fuel systems. It prescribes procedures and designates safe handling, usage, and maintenance of oxygen-fuel systems. Among the topics discussed are the following:

1. Limiting the amount of fuel available when handling or storing oxygen.
2. Proper storage of fuels, including special features of storage areas.
3. Types of equipment to be used in oxygen-fuel systems.
4. Protective equipment and special safety features of equipment used with oxygen-fuel systems.

MONTHLY SAFETY INSPECTION REPORT
COMPRESSED GAS CYLINDERS

1. Are compressed gas cylinders free of dents, cuts, and damaged valve cover threads?
2. Are compressed gas cylinders legibly marked to clearly identify their contents?
3. Are compressed gas cylinders stored in areas which are protected from external heat sources such as open flames, intense radiant heat, electrical arcs, or high-temperature surfaces?
4. Are compressed gas cylinders located or stored in areas where they will not be struck by passing traffic or falling objects?
5. Are compressed gas cylinders protected against corrosion due to moisture and chemical contact?
6. Are compressed gas cylinders effectively secured in the upright position during storage and use?
7. Are compressed gas cylinder valves closed before the cylinders are moved, when the cylinders are empty, and upon completion of each job?
8. Are compressed gas cylinder valve protectors always placed on the cylinders when the cylinders are not in use or connected?
9. Are compressed gas cylinders transported in a manner to prevent creating a hazard by tripping, falling, or rolling?
10. Are compressed gas cylinders of oxygen and fuel stored at least 25 feet apart or separated by a 1/2-hour-rated fire wall at least 5 feet in height?

THE USE OF CHEMICALS

The chemical fire problem affects a wide range of activities, including the use, processing, production, handling, storage, transportation, and disposal of hazardous materials. Unfortunately, there is no single publication or series of publications that could answer every conceivable question about dangerous chemicals. The subject of chemicals is so wide and vast that it requires special in-depth treatment.

More than 3000 chemicals have been classified as dangerous in one respect or another, so it would be impractical to attempt to learn the properties of even several hundred of the most hazardous chemicals. One approach to the problem might be to learn what the dangers are and then determine what group of chemicals have been included in the various classes of hazards. This is not as easy as it appears. Some chemicals may have more than one hazard, all of which can be equally dangerous. A chemical may be more dangerous in one state than in another. As an example, powdered aluminum is more hazardous when damp than when it is either completely wet or dry. There are many ways to group chemicals. The following arrangement has been chosen for this chapter.

Combustible Chemicals

This group comprises those chemicals which are not dangerous or hazardous to health from the standpoint of being violently reactive or explosive. They will burn and yet are more hazardous than charcoal but less hazardous than nitromethane; more hazardous than compressed air, but less hazardous than vinyl chloride. Combustible chemicals would include a wide variety of substances which, in addition to combustibility, are also susceptible to spontaneous ignition (e.g., charcoal and linseed oil).

Oxidizing Chemicals

The oxidizing chemicals are those solids, such as sodium nitrate, which intensify the combustion of other materials but are not as hazardous as other flammable substances often classified as oxidizing materials. Liquid chemicals included in this group are those such as nitric acid, which can cause ignition of combustible materials, but are not as hazardous as perchloric acid and other unstable chemicals. This group also includes gases such as nitrous oxide, which support combustion, but are not as reactive as ozone. Table 4–5 presents the classes of oxidizing materials.

Table 4–5. Classes of Oxidizing Materials

Hazardous Class	Definition	Examples
Oxidizer	A substance that yields O_2 readily to stimulate the combustion of organic matter.	Silver nitrate
Organic Peroxide	An organic derivative of the inorganic compound, hydrogen peroxide.	Lauroyl peroxide
Oxygen	An odorless, colorless, gaseous chemical element that supports combustion. At low temperatures the gas liquefies.	Oxygen

Air-Reactive and Water-Reactive Chemicals

These chemicals may heat spontaneously, ignite, yield toxic or flammable gases, or react violently when exposed to air or water. Included are chemicals such as carbides and phosphides.

Unstable Chemicals

These chemicals (such as acetaldehyde, nitromethane, and others) are likely to decompose or polymerize violently when in an unstable condition, or when involved in a fire. A major hazard connected with these chemicals is the possibility that their containers may rupture and explode.

Another chemical, hydrogen cyanide, is both flammable and poisonous, and the liquid can polymerize explosively. The flash point is 0° F, and the explosive range is from 5.6 to 40%.

Styrene polymerization reaction increases as the temperature increases and eventually the reaction can become very violent unless it is controlled.

EXPLOSIVES AND BLASTING AGENTS

This group includes those chemicals intended to be used as explosives, including blasting agents. Examples are dynamite or smokeless powder. An explosion takes place with a sudden release of a large amount of energy in a destructive manner. No attempt should be made to fight a fire that cannot be contained or controlled before it reaches explosive materials. Table 4–6 contains a listing of explosive classes.

The manufacture, transportation, and possession of explosive materials, is closely regulated by local, state, and federal laws. Anyone conducting an operation or activity requiring the use of explosive materials should review applicable regulations, before obtaining the proper permit from the state marshal's office and/or the U.S. Department of Transportation. When no legal restriction governs the storage of explosives, the recommendations of the Institute of Makers of Explosives should be adhered to. These specify quantities of explosives that may be stored safely at various distances from inhabited buildings, passenger railways, and public highways, as developed by the Institute of Makers of Explosives (*NFPA 495*, *Explosive Materials Code*).

Table 4–6. Classes of Explosives

Hazardous Class	Definition	Examples
Explosive	Any chemical compound, mixture, or device, the purpose of which is to function by explosion, that is, with substantial instantaneous release of gas or heat.	
Class A	A detonating or otherwise maximum hazard.	Black powder, dynamite, blasting caps
Class B	Function by rapid combustion rather than detonation.	Special fireworks, flash powders
Class C	Materials that do not ordinarily detonate in restricted quantities—minimum explosion hazard.	Flares, small arms

Blasting agents should be stored, transported, and used in the same manner as other explosive materials. Blasting agents consist of an oxidizer mixed with fuel and are manufactured so that the final product is relatively insensitive. Because oxidizers are sensitive to heat, impact, friction, and impurities they should be processed and stored in accordance with the manufacturer's recommendations.

CORROSIVES

There are a great many chemicals which will produce serious burns in contact with the skin; some are liquids and others are solids. Some corrosive liquids will react violently with water, while others are combustible liquids such as acetic acid. This acid also gives off vapors that can cause nose, throat, and eye irritation. The classes of corrosives and poisons is shown in Table 4–7.

A corrosive material is a liquid or solid that causes visible destruction or irreversible changes in human skin tissue at the site of contact, or a liquid that has a severe corrosion rate on steel and aluminum. This falls under the *NFPA 471 (Class 8) Recommended Practice for Responding to Hazardous Materials Incidents*. An example of Class 8 materials include nitric acid, phosphorous trichlorine, sulfuric acid, and sodium hydroxide.

Table 4–7. Classes of Corrosives and Poisons

Hazardous Class	Definition	Examples
Corrosive	A liquid or solid that causes visible destruction or irreversible alterations in human skin tissue, or liquid that has a severe corrosion rate on steel.	Sulfuric acid, nitric acid
Poison A	Poisonous gases or liquids—small amount of the gas or vapor of the liquid mixed with air is dangerous to life.	Bromacetone, Cyanogen
Poison B	Poisonous liquid or solid which is known to be toxic to humans, as to afford a hazard to health during transportation.	Potassium arsenate
Irritating	A solid or liquid substance which upon contact with air or when exposed to air gives off dangerous or intensely irritating fumes.	Bromide, tear gas candle
Chlorine	A greenish yellow gas which is an element.	Chlorine

TOXIC CHEMICALS

Many toxic chemicals, such as carbon tetrachloride, have severe neurological effects when inhaled. Some are also toxic through skin absorption such as cresol and methyl bromide. Toxic solutions may be formed when water soluble poisons such as fluorides and cyanide react with water during fire fighting operations when storage containers made of fiberboard are weakened by soaking in water. Some gases and vapors which are highly potent toxins are immediately dangerous to life and health at very low concentrations in air, such as chloropicrin (vapor) and diazomethane (gas).

The Environmental Protection Agency (EPA) uses the term toxic chemical for chemicals whose total emissions or releases must be reported annually by owners and operators of certain facilities and manufacture, process, or otherwise use a listed toxic chemical. Based on a variety of environmental laws mandated by Congress, the U.S. EPA has developed a variety of regulations and policies to implement congressional intent. The list of toxic chemicals is identified in Title III of SARA.

HALOGENS AND HALOGENATED HYDROCARBONS

The ordinary halogens are four chemical elements: fluorine, chlorine, bromine, and iodine. The introduction of a halogen into an organic compound is called halogenation. With metals, the halogens form compounds called halides. Fluorine leaks can cause spontaneous ignition of most materials, while chlorine leaks not only create a poisonous gas hazard but can also corrode unprotected metals as well.

It should be noted that halons remain in the atmosphere for a very long time. The "atmospheric life" of Halons 1211, 1301, and 2402 is 300 years. As these compounds reach the stratosphere (15–50 km above the earth), they release chlorine and bromine. These elements deplete the ozone layer as CFCs do, by combining with one atom of oxygen from ozone (O_3), breaking the ozone apart. Ozone screens out ultraviolet radiation. The "hole" in the ozone layer allows UV radiation to reach the surface of the Earth. Prolonged UV exposure results in melanomas, cataracts, and immune system failure in addition to alteration of aquatic and terrestrial ecosystems.

As a result of these potential risks, in 1987 the United States was one of 24 nations to sign the Montreal Protocol on Substances that Deplete the Ozone Layer. The document called for production and consumption of Halons 1211, 1301, and 2402 to be frozen at 1986 levels. Production and consumption were to be cut in half by 1998. Beginning in 1990—and at least every four years thereafter—on the basis of the latest scientific, technical, and economic information as the evidence of ozone damage accumulated, new chemicals could be added or dropped, additional uses could be banned, and the phaseout schedules tightened accordingly. In the United States, EPA has enacted additional rules regulating the production, use, handling, and depositing of halons. In 1989 the Omnibus Budget Reconciliation Act increased the federal excise tax on halons.

In 1992 NASA concluded that "the rate of destruction to stratospheric ozone may be greater than seriously believed." President George Bush called for the complete phaseout of halons by December 31, 1995. EPA published a list of prohibited substances and acceptable alternatives, making it unlawful to replace halons with substitutes that may present adverse effects to human health or the environment when an alternative has been identified and is currently or potentially available that reduces the overall risks to human health and the environment.

Halons and halon alternatives are discussed in greater detail in Chapter 7.

RADIOACTIVE CHEMICALS

Radioactive chemicals include both naturally occurring chemicals, such as radium, and radioactive isotopes produced from such elements as cobalt. A small quantity of these chemicals creates a significant danger not only of radiation, but in some cases explosions. Classes of radioactive materials are listed in Table 4–8.

Table 4–8. Classes of Radioactive Materials

Hazardous Class	Definition	Examples
Radioactive I	Packages that may be transported in unlimited numbers and in any arrangement, and which require no nuclear criticality safety controls during transportation.	Package containing any radioactive material-measurement of 0.5 millirem or less per hour at each point on the external surface of the transportation package.
Radioactive II	Packages that may be transported together in any arrangment, but in numbers which do not exceed a transport index of 50; no nuclear criticality safety control by the shipper during transportation.	Package containing any radioactive material measuring more than 0.5, but not more than 50, millirems per hour, and not exceeding 1.0 millirem per hour at 3 feet.
Radioactive III	Shipments of packages that do not meet the requirements of Class I or II and which are controlled to provide nuclear criticality safety in transportation.	Package containing any radioactive materials which (1) measures more than 50 millirems per hour at each point or exceeds 1.0 millirems per hour at 3 feet on external surface; (2) contains large quantity as defined by DOT.

DETERMINATION OF HAZARDS

In addition to knowing the chemical properties of materials, a determination of the fire and explosion hazard involves an extensive knowledge of:

1. the nature of raw materials and how these materials are combined (any chemical processing involved, including such operations as oxidation, hydrogenation, nitration, etc.);
2. the physical operations involved, known as unit operations, which include filtration, distillation, etc.;
3. the ability of operating personnel to avoid errors which can lead to unsafe conditions and their ability to effect a safe shutdown when unsafe situations occur (the human element of training and experience);
4. the actual physical plant layout, including site, climatic factors, and protection features.

Table 4–9 summarizes the potential hazards and emergency response actions for explosions and/or fires involving hazardous materials.

Table 4–9. Potential Hazards and Emergency Action for Hazardous Materials

Explosive Class "C" and "B" **Potential** Fire or Explosion—may burn rapidly. Single containers may explode, without causing mass explosion **Emergency Action** Fire—extinguish by conventional methods (water)	**Potential** Fire or Explosion—container may explode in heat of fire; mixtures with fuel may explode; materials may ignite in oxygen vapor **Emergency Action** Fire—SMALL FIRES: Dry chemical or CO_2 LARGE FIRES: Foam or water
Explosive Class "A" Oxidizer **Potential** Fire or Explosion—may explode from heat, flame, or shock **Emergency Action** Fire—extinguish by conventional methods; cannot be properly handled from maximum distance	**Potential** Fire—may ignite combustibles (wood, paper, etc.); reaction with fuels may be violent **Emergency Action** Fire—SMALL FIRES: Dry chemical or CO_2 LARGE FIRES: Water spray or fog
Nonflammable Gas **Potential** Fire or Explosion—container may explode in heat or fire **Emergency Action** Fire—SMALL FIRES: Dry chemical or CO_2 LARGE FIRES: Foam or water	**Oxidizer (corrosive, self-reactive, or thermally unstable)** Fire—SMALL FIRES: Dry chemical or CO_2 LARGE FIRES: Flood with water **Poison, Radioactive Corrosive** Fire—SMALL FIRES: Dry chemical or CO_2 LARGE FIRES: Foam or water
Flammable Gas **Potential** Fire or Explosion—may be ignited by heat, sparks; flammable vapors may spread away from spill; containers may explode in heat or fire **Emergency Action** Fire—SMALL FIRES: Dry chemical or CO_2 LARGE FIRES: Water spray or fog	**Pyrophoric Liquids** **Potential** Fire—may ignite itself if exposed to air, flames; may re-ignite after fire is extinguished; may burn rapidly with flare-burning effect **Emergency Action** Fire—SMALL FIRES: Dry chemical or CO_2 LARGE FIRES: Flood with water
Combustible or Flammable Liquid **Potential** Fire or Explosion—may be ignited by heat, sparks, flames; flammable vapors may spread away from spill **Emergency Action** Fire—SMALL FIRES: Dry chemical or CO_2 LARGE FIRES: Water spray or fog	**Flammable Solid** **Potential** Fire—may be ignited by heat, sparks, flames; may burn rapidly with flare-burning effect **Emergency Action** Fire—SMALL FIRES: Dry chemical or CO_2 LARGE FIRES: Foam or water
Combustible or Flammable Liquid Poison, Self-Reactive or Water-Reactive **Potential** Fire or Explosion—may be ignited by heat, sparks, flames **Emergency Action** Fire—SMALL FIRES: Dry chemical or CO_2 LARGE FIRES: Dry chemical	**Flammable Solid** **Potential** Fire—may ignite itself if exposed to air; contact with water produces flammable gas; may burn rapidly with flare-burning effect **Emergency Action** Fire—DO NOT USE WATER SMALL FIRES: Dry chemical or CO_2

A very real problem today in the chemical, petroleum, and petro-chemical industry where pressures and temperatures are used with highly volatile and flammable liquids is the occurrence of BLEVES. This refers to boiling liquid-expanding vapor explosions. Mainly, this is a large release of energy under explosive conditions (Isman, Carlson, 1980).

COMBUSTIBLE SOLIDS

Many of the principal problems in fire protection arise from the chemical and physical properties of combustible solids. When ordinary wood construction burns, toxic smoke and hot gases are given off; thus, a surface cooling is required before combustion can be halted. If the wood is coated with paint or varnish, a different type of fire control problem results. The problem of dust explosions must be given high priority because of their potential for causing extensive property damage and employee injury.

The only general statement that can be made about the combustion mechanism of common combustible solids is that almost all of them must be heated by some external means, such as a flame or an impinging spark. The process of heating with subsequent burning of solids may occur even in the absence of initiation by a flame or a spark. If a solid is steadily heated by contacting (or being irradiated by) a surface having an elevated temperature sufficiently high to cause decomposition or vaporization of the combustible solid, the exposed surface of the solid may reach its spontaneous ignition temperature.

An important consideration concerning the combustion characteristics of solids is the fact that the ease of ignition and the rate of burning of all combustible solids is governed by their physical shape or their geometric configuration. When a match is applied to wood shavings, ignition will occur almost immediately; if the match is applied to a 2-inch cube of wood, no ignition will take place even after the match is burned to the end. An analysis of this situation reveals two factors that control the speed of ignition and the flame propagation in solids: their heat conductivity and the extent to which each combustible surface is surrounded by air or oxygen so that burning can proceed.

When combustible solids are pulverized into a fine powder and are completely surrounded by air, they form a dust which is highly combustible. Dust explosions are infrequent, but when they do occur, heavy property loss and even employee fatalities and injuries may be encountered. The ignition and intensity of any dust explosion is influenced by or dependent upon: the nature of the materials, the size and shape of the particles, the concentration of the material in the air, ignition temperature, ease of ignition or energy required to initiate ignition of dust cloud, rate of pressure rise of the material, and the maximum pressure developed from the explosion.

Dust explosions can be prevented, or most certainly the impact lessened considerably, by good engineering design, elimination or control of ignition sources, elimination of dust accumulation by ventilation and good housekeeping, and proper design of the structures and equipment to withstand or relieve the force of an explosion.

COMBUSTIBLE METALS

Metals can be toxic or corrosive. No less than 20 metals are radioactive. Most metals have another hazard that is important: they will burn. The flammable potential of 81 different metallic elements varies as widely as would the same number of hydrocarbons.

Perhaps the single most important condition which regulates the combustibility of a metal is its form and shape. Some metals, difficult to ignite in a solid massive form, burn readily as thin sheets or shavings. As the division becomes finer and finer, the ignition temperature of the metal lowers. As the surface area grows, so does the hazard. An industrial establishment which stores or creates metal powders or dusts creates ingredients for an explosion on its premises. Some metal powders can ignite spontaneously in air. When a material does this at ordinary temperatures, it is said to be *pyrophoric*. Some moist metal powders are capable of producing an explosion more violent than one caused by TNT. Many metals will react violently with water in one way or another no matter what their form, massive or finely granulated (Planer, 1979).

Fire fighting methods must be considered carefully when metals, particularly metal powders, are involved. All of the common extinguishers—water, foam, dry chemical, and some of the inert gases—may either stimulate the burning process or cause an explosion. A great deal of thought about alternatives and consequences must go into the size-up and decision, especially if the fire also involves other types of combustibles, such as flammable liquids. The extinguishing agent recommended for a flammable liquid may cause a disaster if it meets a metal powder. (For example, certain metals react exothermically with some inert gases—such as nitrogen—that may be used in extinguishing flammable liquids. So the only acceptable inert gases for these metals are helium and argon.)

PLASTICS

American industry is busily creating and using an enormous group of man-made materials that are also flammable, such as plastics. Defined in simplest terms, a plastic is a material that becomes moldable at least once because of chemical treatment, heat, or pressure, and subsequently hardens into a new shape. Whatever their properties or form, however, most plastics fall into one of two groups—thermoplastics or thermosets (Planer, 1979).

Thermoplastic resins can be repeatedly softened and hardened by heating and cooling, without a chemical change taking place. Thermoplastics usually are in the form of pellets or granules.

Thermoset resins, once polymerized or hardened, cannot be softened by heating without causing a chemical change that would degrade the resin.

Additionally, there are two other classifications used in describing plastics in fire protection today. These groups involve the use of the terms low or high density, with low-density plastics referring to foam or cellulosic foam. The low-density foams are of considerable concern to fire protection personnel as they are being used extensively throughout the construction industry primarily for insulation and also in furniture in the form of upholstery materials. These materials in large quantities should be used with caution because of their flammability and low melting points. Current research is available for treating foam (polyurethane) to be fire retardant.

Although when in a solid massive form many plastics can be difficult to ignite and will not continue to burn on removal of exterior heat source, nearly all will burn rapidly in the form of dust, and if dispersed in the air can be explosively ignited by a spark, flame, or metal surface about 700° F.

Good fire protection starts with the firesafe design of the plant or warehouse and the inspection and modification of the existing facilities, if necessary. Sprinkler-protected, noncombustible construction is appropriate for buildings occupied for storage, processing, and manufacturing of combustibles or flammables, such as those involved in the plastics industry. Automatic sprinklers, standpipe and hose systems, and water-type portable extinguishers should be supplemented by fire extinguishers and special automatic systems suitable for flammable liquid fires and electrical fires, where these hazards exist. Consideration should also be given to the provision of roof vents, particularly in large one-story warehouses of manufacturing plants.

Sprinklers are the most important single system for automatic control of fires in plastic plants. Among the advantages of automatic sprinklers is the fact that they operate directly over the fire and that smoke, toxic gases, and reduced visibility, often associated with fires in plastics, do not affect their operation. Automatic sprinklers, standpipes, and fire hose connections depend upon an adequate water supply delivered with the necessary pressure to control fires.

Sources of Additional Information

Chemical Manufacturers Association
2501 M St. N.W., Washington D.C. 20037
(202) 887-1100

Chemical Transportation Emergency Center (CHEMTREC)
2501 M St. N.W., Washington, D.C. 20037
(800) 262-8200 (800) 424-9300 (Emergency)
(202) 483-7616 (International and Maritime)

Chlorine Institute
2001 L St. N.W., Suite 506, Washington, D.C. 20036
(202) 775-2790

National Agricultural Chemical Association
1155 15th St. N.W., Suite 900, Washington, D.C. 20005
(202) 296-1585

Chemical Institute of Canada
130 Slater St.
Ottawa, Ont. KlP 6E2
(613) 526-4652

REFERENCES

American National Standards Institute. *Responding to Hazardous Material Incidents.* New York: American National Standards Institute, 1993.

American Petroleum Institute. *Venting Atmospheric and Low-Pressure Storage Tanks.* 5th ed. (Std. 2000). Washington, D.C.: American Petroleum Institute, April 1998.

Armour, M. A. *Hazardous Laboratory Chemicals Disposal Guide.* Boca Raton, FL: CRC Press, 1991.

Bahme, Charles W. *Fire Officer's Guide to Dangerous Chemicals.* 2d ed. Boston: National Fire Protection Association, 1978.

Bare, William K. *Introduction to Fire Service and Fire Protection.* New York: John Wiley & Sons, 1978.

Cote, Arthur, and Bugbee, Percy. *Principles of Fire Protection.* Quincy, MA: National Fire Protection Association, 1988.

Donahue, Michael L."CHEMTREC: A Vital Link in Chemical Emergencies," *NFPA Journal* 87 (3):86–90.

Factory Mutual Engineering Corporation. *Handbook of Industrial Loss Prevention.* 2d ed. New York: McGraw-Hill Book Company, 1967.

Hammer, Willie. *Occupational Safety Management and Engineering.* New York: Prentice-Hall, 1989.

Health and Environmental Effects of Oil and Gas Technology. A report to the Federal Interagency Committee on the Health and Environmental Effects of Energy Technology, n.p., n.d.

Isman, Warren E. "Analyze Your Need for a Haz-Mat Team." *NFPA Journal* 85 (5):18.

Isman, Warren E., and Carlson, Gene P. *Hazardous Materials.* Encino, CA: Glencoe Publishing Co., 1980.

McKinnon, Gordon P., general editor. *Industrial Fire Hazards Handbook.* Boston: National Fire Protection Association, 1979.

Mansdorf, S. Z. *Complete Manual of Industrial Safety.* Englewood Cliffs, NJ: Prentice-Hall, 1993.

Meidl, James H. *Flammable Hazardous Materials.* Encino, CA: Glencoe Publishing Co., 1978.

Meyer, Eugene. *Chemistry of Hazardous Materia*l. Englewood Cliffs, NJ: Prentice-Hall, 1977.

National Fire Protection Association. *Fire Protection Handbook.* 17th ed. Quincy, MA: National Fire Protection Association, 1994.

______. *Life Safety Code Handbook.* Quincy, MA: National Fire Protection Association, 1995.

______. *National Fire Codes.* Quincy, MA: National Fire Protection Association, 1995.

______. *National Fire Codes, Standard 30, Flammable and Combustible Liquids Code.* Quincy, MA: National Fire Protection Association, 1983.

______. *National Fire Codes, Standard 58, Liquefied Petroleum Gas Code.* Quincy, MA: National Fire Protection Association, 1998.

______. *National Fire Codes, Standard 704, Identification of the Fire Hazards of Materials.* Quincy, MA: National Fire Protection Association, 1990.

National Safety Council. *Fundamentals of Industrial Hygiene.* 3d ed. Chicago: National Safety Council, 1988.

NIOSH. "Occupational Exposure to Acetylene." *Criteria For A Recommended Standard.* Washington, D.C.: U.S. Department of Health, Education, and Welfare, 1976.

Occupational Safety and Health Administration. *Spray Finishing Using Flammable and Combustible Materials*. Standard Adv. Interpretation 1910.107. Washington, D.C.: Occupational Safety and Health Administration, U.S. Department of Labor, pp. 218–226.

Planer, Robert G. *Fire Loss Control*. New York: Marcel Dekker, 1979.

String, Clyde, and Irvan, Rick. *Emergency Response and Hazardous Chemicals Management*. Delray Beach, FL: 1993, n.p.

Unterberg, W., et al. *How to Respond to Hazardous Chemical Spills*. Park Ridge, NJ: Noyes Data Corporation, 1988.

UPS, Guide for Shipping Hazardous Materials. 5th ed. Published by UPS, 1977.

U.S. Department of Transportation. *1994 Emergency Response Guidebook: Guidebook for Initial Response to Hazardous Materials Incidents*. Chicago: American Labelmark, 1994.

U.S. Office of the Federal Register. *Code of Federal Regulations: 29 Parts 1900 to 1910*. Washington, D.C.: U.S. Government Printing Office, 1993.

______. *Code of Federal Regulations: 49 Parts 100–177*. Washington, D.C.: U.S. Government Printing Office, 1994.

Worobec, Mary Devine, and Hogue, Cheryl. *Toxic Substances Control Guide*. 2d ed. Washington, D.C.: Bureau of National Affairs, 1992.

STUDY GUIDE QUESTIONS

1. What are the major reasons for placing labels and placards on packages and vehicles?
2. Discuss what risks should or should not be taken to identify a hazardous material that is being released.
3. Why is it important to consider isolating or confining flammable liquids? What are some means of accomplishing this?
4. What is considered to be the safest method of handling flammable liquids?
5. Name some sources of ignition.
6. What are oxidizing chemicals? Cryogenic liquids?
7. Explain the role of physical shape or geometric configuration as it relates to the ease of ignition and rate of burning of all combustible solids. Give some examples.
8. Explain the terms flash point and ignition point.
9. What is the *Guidebook for Initial Response to Hazardous Materials Incidents* and what information within it can be useful for fire brigades?
10. What four conditions are indicated by the National Fire Protection Association fire diamond?

Building Construction

CHAPTER 5

CHAPTER CONTENTS

FIGURES

The reports of major fires involving loss of life, heavy property damage, or business interruptions often indicate clear evidence that defects in the original construction or in later additions were major causal factors. Such situations can be addressed by safety professionals with a broad understanding of what fire safety in construction means and how to obtain these goals.

Fire protection in building construction starts at the drawing board, where fire safety errors in the original design can be corrected much more easily and at less cost than can be done after construction is completed. The next action is to make sure that actual construction complies with the code requirements. Finally, completed buildings should be checked for significant changes in the construction or occupancy. If changes have been made, the safety professional must take corrective measures to ensure compliance with fire safety codes and regulations.

To accomplish this, the safety professional must have a working knowledge of fire safety in construction, the tools available to him to achieve this, understand how to use those tools, and understand their limitations.

FACILITY LOCATION

When a facility system is being designed, certainly one of the initial decisions to be made concerns its location. It cannot be overemphasized that the availability and amount of water in reserve is very important when determining a site location. Such information can be attained by contacting local or state public officials and from testing the water system itself.

The quantity of water required is generally determined by the design demands of a protective system utilizing water. NFPA standards contain design guides, flow rates, and storage quantities for water needed in specific systems (*Fire Protection Handbook*, NFPA, 1997). Selection of the water supply source or sources is the next step in planning the location. Potential sources are municipal supply mains, lakes, rivers, on-grade and buried reservoirs, wells, gravity tanks, and pressure tanks.

When available, a municipal water supply is usually the least expensive and most reliable source. A current dependable flow test that records the capability of an available municipal supply is a necessary first step in evaluating the municipal supply source. Water supply information is vital because costly expenditures may be required in the event that the municipal supply is inadequate.

There are several methods of conducting water tests (Planer, 1979). Examples are:

1. Measuring flow from an open butt of a hydrant by attaching a pressure gauge directly to the hydrant. The only accessories being a hydrant cap with a threaded fitting for attaching the pressure gauge by holding a Pilot tube firmly in the center of the water flow just outside the butt or nozzle with the blade at right angles to the nozzle exit.

2. Laying hose from hydrants and measuring flow from nozzles. (When using hose and nozzles, it is necessary to convert the residual pressures at the hydrant to nozzle pressure by considering the flow through the hoses.)

When conducting a water test, the exact diameter of the nozzle or orifice discharging the water must be obtained and accurate gauges must be used. Tests should be made at flows which nearly approximate the facility's water demand.

Pumps can be used to improve a municipal supply when it is deficient in pressure, but they will not improve a municipal supply that is deficient in volume. In may cases the municipal officials will be aware of the deficiencies in their system and may have plans in progress for improvement, or the economic impact of a new facility might provide incentive to the municipality to provide water improvement.

When the municipal water system is found to be inadequate, then a secondary water supply will be required. Lakes and rivers, when available, should be evaluated for continuity of supply and for water quality. Reservoirs are another way to enhance the municipal supply. Once sized, they are developed by pumping. Available reservoir variations include buried steel or concrete; on-grade steel; or concrete and on-grade, plastic-lined, earth-formed. In some instances, the reliability of a reservoir water source may be developed by dividing or otherwise partitioning the source, and sometimes this will serve to provide the equivalent of two water sources. Wells are sometimes developed as a water supply for a vertical fire pump.

Gravity and pressure tanks are also very reliable. They are practically limited to an installed height of about 150 feet above grade or 50 feet above a building. Thus, they are not used when high pressure is needed.

Evaluating the water supply is the primary input of the safety professional in site selection. However, there are other factors of concern which involve the safety professional in the final site selection. These factors are:

- Type of fire protection
- Hazardous exposures
- Probability of flooding
- Earthquake likelihood

LAYOUT OF FACILITIES

The layout and arrangement of a plant or industrial undertaking is an important factor to be considered in the safety function. If the maximum efficiency concerning safety is to be reached and maintained, it must first be planned into the organization's physical layout. An excellent illustration of this is made through a well-accepted definition of good plant layout as: (1) placing the right equipment, (2) coupled with the right method, (3) in the right place, (4) to permit the processing of a product unit in the most effective manner, (5) through the shortest possible distance, (6) in the shortest possible time.

Planning

There are several factors to be considered in planning a facility layout. Some of these factors are: location; spacing and arrangement of power plants, process units, tanks, and other structures; the products that are to be made; the processes that are to be used in making the products; the size and shape of the buildings; the kinds of machinery required; and the size, or approximate size, of the workforce. It should be noted that failure to recognize these factors as well as other factors present in pertinent industries, will result in economic losses sustained by an organization due to changes in construction before actual construction is completed and once construction is complete. The point that should be stressed is that changes in the blueprint stage are relatively cheap as compared to changes made during or after construction (Smith, Harmathy—ASTM).

During the planning stage, it is essential that considerations for fire safety be included. Also during the planning stage, it is vitally important that the safety professional draft a list of items pertinent to his or her type of industry that should be checked against any such plans or blueprints. This is important in case any items or design considerations are excluded in the initial planning process. In such a list, there will be a number of major items, each of which will include many minor items. An example of a list of major items is as follows:

- Site
- Transportation facilities
- Personal Service facilities
- Walkway surfaces
- Lighting, heating, and ventilation
- Elevators
- Boilers and pressure vessels
- Electrical wiring
- Fixed machinery and equipment
- Portable equipment
- Provisions for servicing plant and equipment
- Fire prevention and protection
- Provisions for health and safety

Another approach used by some safety professionals is to draft a list of possible exits, hazardous areas, and vertical openings that may exist in the industry or plant. This list of possible fire hazards may then be examined to determine the design flaws made during the initial planning steps (Marchant).

Floor Design

Fire protection of building elements is provided for two reasons. The first is to limit the spread of fire within a building or prevent the spread into the building during a fire exposure, and the second is to ensure that, even under that exposure, the building frame or elements of that frame will not collapse for a reasonable period of time. Such collapse or even the threat of collapse will render fire fighting measures less effective than they might be otherwise.

It should be recognized, though, that some building designs, such as steel frame and sided buildings, have essentially no fire resistance.This lack of fire resistance is intentional, and will likely result in early collapse of the structure during a fire.

There are two groups of building elements: (1) load bearing, and (2) nonload bearing. Load-bearing elements are those that support loads other than their own weight. Nonload-bearing elements would have no effect on the structural behavior of the building as a whole.

Building codes provide requirements for both structural loads and superimposed live loads. The code requirements are such that building "failures" are rare. Structural failures are generally the result of application of unanticipated loadings. In a fire situation, "loads" are induced by heat which may cause thermal stresses. These stresses may be increased if the members are in any way restrained against expansion. Also, heat may cause a loss of strength, if not actual consumption, of the structural member.

Flow Sheets

Flow sheets are just as they sound, they are diagrams of the actual flow of operations and processes in an organization. A detailed flow sheet is an extremely useful guide in laying out plants, particularly those using dangerous materials and complicated processes. A flow sheet made in sufficient detail to include hazard points and built-in provisions, with the types of potential fires in mind, can be a useful tool used for planning. A flow sheet should be drawn to show the plant layout plan for the entire facility, and should show relationships between buildings and structures, roadways, water lines, services lines, traffic flow, and the location of bulk storage facilities for hazardous substances. In some cases, where it is necessary to bring attention to special procedures or hazards, supplemental sheets (such as building blueprints) may be included with the flow sheet.

The nature of the materials and processes in each manufacturing stage can be studied through the flow sheet and provisions made to control or eliminate fire hazards.

LIFE SAFETY

Assuring the life safety of workers in the event of a fire or other emergency is a paramount responsibility of employers. Fires and other emergencies occur with regular frequency in the workplace. Employers must be prepared by ensuring that employees are warned when an emergency is discovered and can safely escape from a building. Codes and standards have been prepared and adopted by national standards organizations, all levels of government, and insurers. These standards outline expectations concerning life safety ranging from generalized performance standards to more rigorous specification standards. In either case, assuring that life safety is achieved in a facility is a difficult task for the employer. This discussion will identify the relevant life safety standards, explain their fundamental concepts, and describe management approaches that can be implemented to assure that life safety is achieved on an ongoing basis.

Assuring life safety should be a paramount safety concern for any employer. Fires and other emergencies can occur at any time and in any part of a facility. Fires, especially in industrial occupancies, have a high likelihood of occurring. Once more, fires can quickly become severe. A small fire can rapidly progress into a major catastrophe within a few minutes. Large numbers of occupants can become exposed. Therefore, assuring that the layout and contents of a facility are configured to allow for rapid and unobstructed egress from the fire or emergency area is a paramount safety concern.

On September 3, 1991, twenty-five workers died and fifty-six were injured in a chicken processing plant. They could not escape a fire that quickly involved a large portion of the plant within two minutes of igniting. Among the many factors that contributed to this tragedy were locked exit doors and unmarked escape routes. The employer was held criminally responsible for not assuring the life safety of workers in the plant (Klem, 1992). The personal liability suits and potential damages against the employer are immense. A basic fire safety awareness and a small investment in planning by the employer could have prevented this outcome. This incident triggered a public outcry for better fire safety awareness, as well as improved government enforcement.

Assuring life safety is also important to government agencies and insurers. The following organizations regulate life safety in the workplace.

- The Occupational Safety and Health Administration (OSHA) enforces standards in the Code of Federal Regulations (CFR).
- State and local authorities having jurisdiction (AHJ) enforce model building codes, fire codes, and/or a life safety code adopted by the individual jurisdiction.
- Insurers require their insured to abide by nationally recognized standards and government regulations.
- The courts influence employers' actions by awarding punitive and compensatory damages for regulatory violations and personal liability judgments.

Assuring life safety is both a legal and moral responsibility of employers.

Several standards exist related to life safety. *NFPA 101, Life Safety Code*, is prepared and published by the National Fire Protection Association (NFPA). This consensus standard is the most widely referenced and adopted standard pertaining to life safety. The *Life Safety Code* contains fundamental life safety requirements; classifies occupancies; details requirements for means of egress, features of fire protection, building services, and fire protection equipment; and provides requirements for specific types of occupancies (NFPA, 1997). The *Life Safety Code* is written so that it can be adopted by state and local governments for regulatory purposes. It is reviewed, revised, updated, and re-issued every three years by the NFPA.

Life safety requirements can also be found in the model building codes used throughout the United States. Model building codes identify both performance-based and specification requirements for means of egress, fire protection features, and construction features. Model building codes are typically reviewed, revised, and/or updated on a three-year cycle. Four model building codes are primarily utilized in the United States. Figure 5–1 identifies the four model building codes and their sponsoring organizations.

Title	Sponsoring Organization
National Building Code	Building Officials and Code Administrators (BOCA)
Standard Building Code	Southern Building Code Congress, International (SBCCI)
Uniform Building Code	International Conference of Building Officials (ICBO)
One- and Two-Family Dwelling Code	Council of American Building Officials (CABO)

Figure 5–1. Model Building Codes

In the United States nearly every jurisdiction has applicable standards and codes for fire safety. To develop a fire safety plan successfully, knowledge and background of the subject, as well as resourcefulness, are necessary. However, buildings are different in design and structure. In researching the literature the *NFPA 101, Life Safety Code*, is the only code that has specific chapters covering both new and existing buildings.

States and local governments typically adopt a model building code to enforce in their respective jurisdictions. A combination of enforcement methods are used and can include: a building permit process, an occupancy permit process, building plans review, and on-site inspections. These methods bring life safety to the grass roots level. They provide some assurance to the public that buildings are fire safe and that a degree of life safety is achieved.

OSHA regulates life safety in the workplace through its own standards contained in 29 CFR 1910, specifically, 1910.35-37. OSHA incorporated portions of the 1970 *Life Safety Code* by reference, which were not fully adopted until 1980. The employer utilizing OSHA's life safety standards should do so with caution for two reasons. First, OSHA only adopted portions of the 1970 *Life Safety Code*, specifically, performance-based criteria that are open to interpretation. Few details are provided to the employer regarding the layout and configuration of a facility or its contents. Second, the *Life Safety Code* has undergone eight revisions since 1970. This suggests that more current information has been included in recent editions. For example, the 1970 *Life Safety Code* permits a means-of-egress travel distance of 100 feet in an industrial occupancy. The 1994 edition permits a travel distance of 200 feet (Carson, 1993).

OSHA now has a policy that reads: "compliance with an applicable NFPA standard will be considered to be one means of compliance with the performance criteria in the OSHA standard" (Carson, 1993). Safety managers are encouraged to use the most recent edition of the *Life Safety Code* as the basis for assuring life safety in their facilities. Complying with the *Life Safety Code* (or any other NFPA standard) will be considered compliance with the equivalent OSHA standard.

The Americans With Disabilities Act (ADA) also contains life safety requirements. The ADA requires that facilities be accessible to the disabled, as well as providing for adequate egress during emergencies. When considering life safety assurance, the ADA contains a major flaw. It is enforced after the fact. The ADA is a civil rights law and is not linked to model building codes adopted by states and local jurisdictions. This means that a company could construct a building ignoring the ADA's provisions. The company could then be sued after the building is constructed for not incorporating the ADA's provisions. This would provide no life safety assurance to the disabled in the meantime. Recognizing this flaw, the NFPA incorporated egress provisions for the disabled in its 1994 edition of the *Life Safety Code* (Cummings and Jaeger, 1993). This is another justification for utilizing the most recent edition of the *Life Safety Code* as the basis for assuring life safety.

EVALUATING LIFE SAFETY

Every facility is unique in some manner. A *one-size-fits-all* life safety inspection checklist could not meet the needs of every facility. However, armed with a basic knowledge of life safety concepts, an employer, its supervisors, and its employees can identify and abate many common types of life safety hazards. Through education and training, workers are more likely to recognize life safety hazards and less likely to create them in the first place.

Life safety hazards can also be identified by conducting regular inspections. Two types of inspections are recommended for facilities. First, a self-inspection program should be developed and implemented. This encompasses supervisors and employees regularly inspecting their work areas for life safety hazards. Self-inspection programs are effective because supervisors and employees are responsible for assuring life safety in their work areas. They can see the results of their efforts which can increase motivation, pride, and productivity. Next, management should conduct quarterly life safety inspections of the facility. During quarterly inspections, management can confirm that the self-inspection program is achieving results. It is a method to hold supervisors accountable for life safety assurance in their work areas. Both types of inspections should analyze the factors that influence life safety assurance.

The following accepted terms and requirements are provided for those unfamiliar with *Life Safety Code* requirements.

- **Means of Egress.** A continuous and unobstructed way of travel from any point in a building to the public way must be maintained. It has three parts: the exit access, the exit, and the exit discharge. All three parts should be properly configured and maintained.

- **Exit Access.** An exit access should be protected by construction with a 1- to 2-hour fire rating depending on the how many stories a building is.
- **Exit Discharge.** An exit discharge must terminate in a public way.
- **Number of Means of Egress.** All buildings must contain at least two separate means of egress. Each means of egress should be as far remote as possible from the other. Buildings with an occupant load of 500–1000 must have three means of egress and an occupant load greater than 1000 requires at least four means of egress.
- **Egress Width.** The means of egress must maintain a minimum clear width. The minimum clear width permitted is 28″ for existing buildings and 36″ for new buildings.
- **Door Width.** Exit doors must maintain a minimum door opening. Door openings must be 32″ for new buildings, 28″ for existing buildings, and a single door cannot exceed 48″ in width.
- **Impediments.** Impediments that obstruct a means of egress, prevent passage, or reduce passage are not permitted.
- **Locked Doors.** All exit doors must be kept unlocked while a building is occupied. Where security is a concern, panic hardware and other special devices are available to make an exit door open only from the direction of egress.
- **Occupant Load.** The number of occupants permitted in a building at one time should be closely monitored. Occupant load is determined by multiplying the building's gross or net area by an occupant density. The permitted occupant load should be conspicuously posted in a building. It is intended to prevent overcrowding and panic in case of a fire.
- **Exit Capacity.** The total exit capacity of a building must exceed the building's occupant load. Exit capacities are determined for specific occupancies. It is determined by multiplying the exit allowance (persons per inch) by the width of an exit.
- **Travel Distance.** Maximum travel distances are regulated by occupancy type and the installation of automatic sprinkler systems. They should not be exceeded.
- **Lighting.** A means of egress must be illuminated so occupants can safely evacuate.
- **Emergency Lighting.** A redundant means of illuminating the means of egress must be provided if the means of egress will be left in total or partial darkness upon failure of that primary illumination source. Emergency lighting must be provided for ninety minutes following the activation of a fire alarm and should be tested regularly.
- **Marking the Means of Egress.** The means of egress must be marked by approved signs. Any portion of the means of egress that changes direction must be clearly marked. Exit signs must be illuminated.

- **Stairs.** Two protected stairwells should be provided in multistory buildings. Risers, treads, and stair slope must maintain minimum specification. Stairs must not be obstructed and storage is not permitted on or under stairs. In some cases, stairwell pressurization is required to create a pressure differential that prevents smoke from entering a stairwell.

- **Hazard Contents.** A building's contents are classified according to combustibility and smoke production. Hazard contents are regulated for certain occupancies usually based on expected occupant loads, occupant egress ability, and fire protection features.

- **Occupancy.** Buildings and portions of buildings are categorized according to what the building is used for. Specific provisions are required for certain occupancies.

- **Fire Alarm System.** All buildings must have an approved automatic fire alarm system to warn occupants. These systems must be properly designed, installed, and maintained.

- **Compartmentation.** Verifying the integrity of fire barriers such as fire-rated walls, separation walls, partitions, and floor construction is important. These fire barriers subdivide building spaces into smaller areas. Should a fire initiate, the fire-rated barriers are intended to limit fire spread to a single compartment. Penetrations in fire barriers must be protected.

- **Interior Finish.** The materials that comprise the exposed interior surfaces of a building are regulated. Interior finishes that will propagate rapidly spreading fires or produce excessive smoke must be avoided.

- **Headroom.** Headroom for occupants must be maintained. The minimum permitted headroom is 7′6″ and projections from the ceiling must leave a headroom of 6′8″.

When inspections are conducted it is necessary to insure that the required life safety features are present and properly maintained, as well. Safety inspectors will need a thorough understanding of the concepts and requirements of *NFPA 101, Life Safety Code*, to conduct a thorough inspection.

Figure 5–2 will show the unusual condition you will identify and the specific observations that you will have to make when inspecting special structures. Each inspection is worthy of your best professional expertise and judgment.

MANAGEMENT APPROACHES FOR ASSURING LIFE SAFETY

So how does an employer assure that life safety standards are followed? Several management approaches are available. However, combining several management approaches into a process is recommended. A management process for assuring overall life safety in a facility is outlined below.

1. A qualified person conducts a baseline life safety inspection that assures:
   - the reliability of building construction features as fire barriers
   - the number of exits provided is adequate for the occupant load
   - the exit capacity exceeds the occupant load
   - the condition and adequacy of the means of egress and its protective features
   - building contents and interior finishes will not create an excessive fire load
   - the readiness, operability, and maintenance program of fire detection, fire alarm, fire suppression, and smoke management systems
   - that panic conditions will be avoided in the event of a fire
   - the effectiveness of warning procedures and an emergency action plan

Deficiencies should be noted in a written evaluation.

2. Prioritize deficiencies according to the risk to occupants.
3. Establish corrective actions that mitigate deficiencies with both engineering and administrative controls permitted by the *Life Safety Code*.
4. Assign corrective actions to supervisors and establish measurable milestones for implementation.
5. Educate and train workers about how to identify, report, and correct life safety hazards in the facility.
6. Implement a self-inspection program that requires supervisors to regularly inspect their work areas for life safety hazards.

This process will help assure that life safety hazards are identified and mitigated.

BUILDING CODES

Local fire safety codes provide minimum standards. Building codes are concerned with structures to be built. After construction, a building may never come to the attention of local building inspectors. Codes may not deal with complex design considerations of the following:

1. Shopping Malls contain stores of various sizes with goods of varying combustibility, all under one roof, and each opening into a covered mall. Egress is usually by way of the covered mall.
2. Industrial Buildings can be large and undivided. Dividing the buildings into compartments may be impractical and distance to exits is likely to be extreme.
3. Schools with open and flexible layouts, "commons" areas, are unsuitable for compartmentation and access to exits may be varied and obstructed.
4. Membrane Rooted Structures (i.e., domed stadiums) require the quick evacuation of large numbers of people.

Property Name: Owner:
Address: Phone Number:

OCCUPANCY
Occupancy Classification: Change from Last Inspection: Yes No
Occupant Load: Egress Capacity: Any Renovations: Yes No
Vehicle or Vessel: Windowless High Rise Water Surrounded Pier Underground Other

BUILDING SERVICES
Electricity Gas Water Other Are Utilities in Good Working Order: Yes No
Emergency Generator: Yes No Size: Last Date Tested:
Date of Last Full Load Test: In Automatic Position: Yes No
Fire Pump: Yes No GPM: Suction Pressure: System Pressure:
Date Last Tested: Date of Last Flow Test:
In Automatic Position: Yes No Jockey Pump: Yes No

EMERGENCY LIGHTS
Operable: Yes No Tested Monthly: Yes No
Properly Illuminated Egress Paths: Yes No In Good Condition: Yes No

EXIT SIGNS
Illuminated: Internally Externally Emergency Power: Yes No Readily Visible: Yes No

FIRE ALARM: Yes No Location of Panel:
Coverage: Building Partial Monitored: Yes No Method:
Type of Initiation Devices: Smoke Heat Manual Water Flow Special Systems
Date of Last Test: Date of Last Inspection:
Notification Signal Adequate: Yes No Fire Department Notification: Yes No

FIRE EXTINGUISHERS
Proper Type for Hazard Protecting: Yes No Mounted Properly: Yes No
Date of Last Inspection: Adequate Number: Yes No

FIRE PROTECTION SYSTEMS
Type: Sprinkler Halon CO_2 Standpipe Water Spray Foam Dry Chemical Wet Chemical Other
Coverage: Building Partial Date of Last Inspection:
Cylinder or Gauge Pressure(s): 1 psi., 2 psi., 3 psi., 4 psi., 5 psi.
Valves Supervised: Electrical Lock Seal Other Are Valves Accessible: Yes No
System Operational: Yes No Sprinkler Heads 18 in. from Storage: Yes No

FIRE RESISTIVE (FR) CONSTRUCTION
Stairway FR: Yes No Hourly Rating: Corridors FR: Yes No Hourly Rating:
Elevator Shaft FR: Yes No Hourly Rating: Floor-Ceiling Assemblies FR: Yes No Hourly Rating:
Major Structural Members FR: Yes No Hourly Rating:
Floor-Ceiling Assemblies FR: Yes No Hourly Rating:
All Openings Protected in FR Walls and Floor-Ceiling Assemblies: Yes No

HAZARDOUS AREAS
Protected by: Fire-Rated Separation Extinguishing System Both Door Self-Closures: Yes No

HOUSEKEEPING
Areas Free of Excessive Combustibles: Yes No Smoking Regulated: Yes No

INTERIOR FINISH
Walls and Ceilings Proper Rating: Yes No Floor Finish Proper Rating: Yes No

MEANS OF EGRESS
Readily Visible: Yes No Clear and Unobstructed: Yes No
Two Remote Exits Available: Yes No Travel Distance within Limits: Yes No
Common Path of Travel within Limits: Yes No Dead-Ends within Limits: Yes No
50% Maximum through Level of Exit Discharge: Yes No
Adequate Illumination: Yes No Proper Rating of All Components: Yes No
All Exit Enclosures Free of Storage: Yes No
Door Swing in the Direction of Egress Travel (when required): Yes No
Panic/Fire Exit Hardware Where Required: Yes No Operable: Yes No
Doors Open Easily: Yes No Self-Closures Operable: Yes No
Doors Closed or Held Open with Automatic Closures: Yes No
Corridors and Aisles of Sufficient Size: Yes No Stairwell Re-Entry: Yes No
Mezzanines: Yes No Proper Exits: Yes No

VERTICAL OPENINGS
Properly Protected: Yes No Atrium: Yes No Properly Protected: Yes No
Are Fire Doors in Good Working Order: Yes No

HIGH RISE
Central Control Station
Voice Fire Alarm Panel: Yes No
Fire Department Two Way Telephone Communications: Yes No
Fire Alarm System Annunciators: Yes No Elevator Floor and Control Annunciators: Yes No
Emergency Generator Status Annunciator: Yes No
Fire Pump Status Annunciator: Yes No
Telephone for Fire Department Use: Yes No

EMERGENCY POWER
Emergency Lights: Yes No Fire Alarm System: Yes No
Electric Fire Pump: Yes No
Central Control Station Equipment and Lighting: Yes No
At Least 1 Elevator Serving All Floors: Yes No Transferable: Yes No
Mechanical Equipment for Smoke Control: Yes No

OPERATING FEATURES
Fire Drills Conducted: Yes No Employees Trained: Yes No

Figure 5–2. Special Structure Occupancy Fire Inspection Form

5. Hotel Exhibition Halls involve heavy fire loading and contain numerous ignition sources.
6. Windowless Buildings make entrance for fire fighting and rescue difficult.
7. High-Rise Buildings with mixed occupancy present planning problems and make rescue difficult. Local fire departments may not have the resources necessary to effectively control high-rise fires.

Local codes require that all buildings be structurally stable. Buildings of certain sizes are required to be *fire resistive*—that they be able to withstand collapse during fires. Brannigan (1982) comments, "If a building is not required by law to be fire resistive, the designer is *not required* to give any consideration to its potential collapse in a fire (emphasis added). Some nonfire-resistive elements, such as sawn wooden 'I' beams, have historically demonstrated certain resistance to collapse. If a builder chooses to substitute elements of literally no fire resistance, such as fabricated 'I' beams, which will carry normal load, the codes do not object." To insure containment of fire to the building of origin, local codes include:

- Regulation of wood shingles
- Requirements for masonry exterior walls
- Distance between buildings, or alternatives (i.e., fire-resistive exterior walls, rolling shutters, exterior sprinklers)

To contain the fire to specified areas within the building, local codes include:

- Gypsum board sheathing in combustible buildings
- Fire-resistive floors
- Fire walls to subdivide floor areas
- Closure devices on openings in barriers (i.e., fire doors, enclosed stairways with self-closing doors)

To insure safe egress of occupants, local codes include:

- Outside fire escapes
- Enclosed interior stairways
- Exitways based on the number of occupants permitted
- Limited flame spread in corridors (Cote and Bugbee 1988, Brannigan 1982)

The safety manager must be particularly astute in facility design. Building codes are guidelines for confining fire to a *manageable* (whatever that might be) area where it can be extinguished by fire fighters, and for enabling the safe evacuation of the building's occupants. Brannigan (1982) offered the following comment on the deficiencies of building codes.

> Even the best building codes can have technical deficiencies. Provisions are handed down from code to code, often without any valid basis. Others are based on the influence

exerted on code-making authorities by proponents of one material or another. Building code provisions are voted on in meetings at various levels and then by the appropriate political body. All these votes are subject to the political process in all its ramifications. It is not inaccurate to state that building codes are political rather than technical documents. It is disturbing to hear a fire officer say, "This building can't have any problems. It was built to the latest code."

The most important consideration in building design is the providing of exits based on the number and characteristics of occupants, the maintenance of exits, and the ability of occupants to reach exits quickly.

When designing a system of egress, it is necessary to consider the physical dimensions of a building's occupants. From this safety engineers have created the *body ellipse*. The majority of adult males do not exceed a width of 20.7 inches (52.8 cm) at the shoulders. For Americans the body ellipse is given a major axis of 24 inches (60.9 cm) and a minor axis of 18 inches (45.7 cm). The additional width added is a result of clothing in colder climates and a propensity of Americans to establish a buffer zone between themselves and others. The body ellipse has an area of 2.3 square feet (0.21 square meters) and represents the area of a standing person (Cote and Bugbee 1988). While walking, lateral movement or *swaying* occurs. The sway can vary, as a result of crowding or moving on stairs, from 1.5 inches (3.8 cm) to 4 inches (10.2 cm). A total width of approximately 30 inches (76.2 cm) is needed for a single file of people moving up or down stairs. As crowding increases, movement decreases. Walking speed slows to a shuffle and finally stops.

Success in fire safety management depends on an effort of managers and personnel to identify and correct fire hazards. As might be suspected, when a fire occurs safety consists of having the occupants leave the building before conditions become life threatening. Cote and Bugbee (1988) stress several requirements for evacuations to work safely.

- The fire is detected and the alarm is promptly activated.
- The occupants recognize the alarm signal.
- The occupants immediately proceed to evacuate the building.
- The occupants proceed toward the exits in an orderly and efficient manner.
- The means of egress are adequate to accommodate the number of occupants.
- The exits have been properly designed, constructed, and maintained to provide a safe environment.
- The last occupant is out of the building before the fire develops life-threatening conditions.

In a perfect world, all occupants would be successfully evacuated before life-threatening conditions developed. Certain facilities (e.g., hospitals and jails) do not present unrestricted means of egress. It may be necessary to defend the facility to provide protection for its occupants. A defensive fire safety plan contains these points:

- The size of any expected fire is controlled. This is accomplished by controlling the combustibility of interior finish and furnishing to limit the speed with which a fire would develop and spread. Automatic sprinklers normally are considered necessary to assure that any fires that do start are quickly controlled.

- The building is compartmented. Dividing the building with fire barriers limits the number of people exposed to a fire. Those in the area exposed to the fire can move horizontally through a fire barrier to an area of refuge.
- Exits are provided. Adequate exits must be maintained in case building evacuation is required. Evacuation is generally a last resort (Cote and Bugbee 1988).

Building codes contain minimum requirements for design and construction of buildings. Specification codes describe what materials can be used in construction, the building size, and how to assemble the structure's components. Performance codes contain objectives and criteria for satisfying those objectives. Remember that building codes contain minimum requirements. In the real world an optimum level of safety is weighed against cost. Alternate materials and methods may be substituted if they can be shown adequate under minimum code requirements. Minimum code requirements include specifications for structural stability, fire resistance, means of egress, sanitation, lighting, ventilation, and safety equipment (e.g., alarms and fire suppression systems).

Because building codes are laws, many local and state jurisdictions write their own building codes. Municipalities may have difficulty in distinguishing what should be contained in a building code, a *Life Safety Code*, or a fire prevention code. Model building codes are developed by three organizations and are usually incorporated by reference into local and state statutes.

Building Officials and Code Administrators (BOCA) publishes the BOCA *National Building Code*, issued every three years. It has been incorporated into local codes primarily in the Midwest, New England, and Mid-Atlantic states. The BOCA *National Fire Prevention Code* is a supplement designed to eliminate conflicts with building code requirements. The International Conference of Building Officials (ICBO) publishes the *Uniform Building Code* and the *Fire Prevention Code* (issued every three years with supplements issued annually). It has been incorporated into local codes primarily in the Pacific and Mountain states. The Southern Building Code Congress International (SBCCI) publishes the *Standard Building Code* (issued every three years with supplements issued annually). It has been incorporated into local codes in the South.

The content of the building code is important to safety managers when considering plant design and layout. Adequate space, light, and ventilation enable workers to efficiently perform work and maintenance tasks, and to safely store equipment and supplies. As management style moves toward mutually agreed upon decisions which allow the organization to operate successfully, those responsible for fire protection can provide input to those responsible for building design.

A facility's location will determine its design. The response time of emergency service providers (including the fire department) is an important factor in design. Traffic congestion and limited-access highways may slow response time. Building design can prevent the facility from being easily approached by fire apparatus. In rural and congested urban areas the number, location, and spacing of water mains and hydrants may be inadequate for fire apparatus, sprinklers, and standpipes.

ENGINEERING DESIGN CONSIDERATIONS FOR PLANT LAYOUT

When planning the layout of a plant, several engineering design considerations should be taken into account. If the planner is familiar with existing conditions and applies fire prevention principles, he will not overlook the following:

- **Adequate Space.** Adequate space must be provided for the completion of the employees' jobs. Also adequate space should be provided for storage of equipment and supplies.
- **Safe Access.** Safe access must be provided for an employee to successfully fulfill his job requirements. Safe access must also be provided for an employee whereever his job requires him to go.
- **Safe Maintenance.** Safe working conditions should be provided to employees performing maintenance functions.
- **Adequate Air and Light.** Adequate air and light must be provided to employees so that they may carry out the functions of their respective jobs.
- **Services.** Provisions are made for the arrangement of machinery and equipment and provisions must also be made for the servicing of those areas.
- **Expansion.** A far-sighted planner allows for expansion. If this is properly done, the need for rearranging departments, at a later date, to overcome congestion is avoided.

LOCATION OF BUILDINGS AND STRUCTURES

One of the basic fundamentals of plant layout is segregation. The segregation can take place in various ways, but the most common is by raw materials storage, processing buildings, and storage of finished materials. In planning the location of various units in a plant, the codes of the National Fire Protection Association and of local and state authorities should be followed. Some general rules that should be followed in the layout of buildings and structures are as follows:

- Ample space would be provided between segregated units and flame sources. Plants using and storing flammable materials should conform to the specifications developed by the *Flammable and Combustible Liquids Codes, National Fire Codes, No. 30.*
- Types of retardants used in relation to the distance between buildings of frame, brick, and fire-resistant construction should conform to the *National Fire Protection Association Code, No. 80A, Protection of Buildings from Exterior Fire Exposures.*

- Power plants and employee amenity areas should be placed where they have minimum exposure to the operating unit.
- Loading and filling installations of such gases as propane, should be located several hundred feet away from ignition sources. *NFPA 58* requires distances and requirements for propane. Propane must be emphasized as a dangerous gas.
- Processing units should be grouped according to the type of materials being handled.

Where no legal restriction governs the storage of explosives the recommendations of the Institute of Makers of Explosives should be followed. This specifies the quantities of explosives that may be stored safely at various distances from inhabited buildings, passenger railways, and public highways, as developed by the Institute of Makers of Explosives (*NFPA 495, Explosive Materials Code*).

SELECTION OF BUILDING MATERIALS

Fire protection begins on the drafting board where full consideration should be given to heights, areas, egresses, construction and finish materials, structural assemblies, occupancy factors, private and public fire protection, exposure protection, and limitations imposed by building codes, state fire codes, and other legal requirements. All of these factors are interrelated in the building construction. Approximately 75 to 80% of building codes involve fire safety requirements (Egan, 1978). Many buildings reflect the practice of design without regard for fire protection.

Selecting the appropriate materials and assemblies to meet expected fire conditions in a structure requires familiarity with fire properties of building materials and structural assemblies. These properties are divided into two basic categories: combustibility of materials and fire resistancy of materials. Combustibility is further divided into three categories: flame spread, fuel contributed, and smoke development.

- **Brick** is quite fire resistant, but brick or mortar may deteriorate.
- **Stone** of many types may spall (i.e., lose part of its surface when heated).
- **Cast Iron** has good fire characteristics. The casting method determines whether it is good or bad. This cannot be determined by examination. Poor connections (rivets or welds) are the chief cause of failure.
- **Steel** elongates substantially at about 1000° F. If restrained, it will buckle. It fails at about 1300° F. Water doesn't cause failure. It cools steel. Cold drawn steel (e.g., cables and wire ropes) fails at 800° F.
- **Reinforced Concrete** is a composite material. Failure of the bond between concrete, which provides compressive strength, and steel, which provides tensile strength, causes failures of reinforced members. Concrete may be formulated to perform its intended function under fire conditions within certain limits.

- **Gypsum** absorbs heat from the fire as it breaks down under fire exposure.
- **Plastics** have a wide variety of properties. Some ignite and burn readily. Others do not. Some burn with heavy smoke production. Others do not (Brannigan, 1982).

Local fire departments and building inspectors, the state fire marshal's office, and building contractors can provide assistance to safety professionals in building design and construction. Our objective is to reduce losses from fire by making construction decisions based on standards of prudent design and selection of materials.

The type of construction selected is dictated by the type of occupancy, expected occupancy hazards, and threats from exposure hazards. There are five types of building construction (*NFPA 220*) from which to make a selection. These are:

- **Type I** - Construction is that type in which the structural members, including walls, columns, beams, floors, and roofs, are of approved noncombustible or limited-combustible materials and have fire resistance ratings.
- **Type II** - Construction is that type not qualifying as Type I Construction in which walls, beams, columns, floors, and roofs are of approved noncombustible or limited-combustible material and have fire resistance ratings.
- **Type III** - Construction is that type in which exterior walls and structural members which are portions of exterior walls are of approved noncombustible or limited-combustible materials and interior structural members, including walls, columns, beams, floors, and roofs, are wholly or partly of wood of smaller dimensions than required for Type IV Construction or of approved noncombustible materials, limited combustibles, or other approved combustible materials.
- **Type IV** - Construction is the type in which exterior and interior walls and structural members, which are portions of such walls, are of approved noncombustible or limited-combustible materials.
- **Type V** - Construction is that type in which exterior walls, bearing walls, floors, and roofs and their supports are wholly or partly of wood or other approved combustible materials smaller than required for Type IV Construction.

In addition, all five types of construction must have structural members that have fire resistance ratings not less than those set forth in NFPA's Table 3 or Fire Resistance Requirements. The table also provides information relative to some exceptions. There are many combinations of these general types.

A building is composed of various types of structural members that are arranged and joined together so as to support a given load or loads, or to shelter. The various components that make up the building are (1) the general framing (foundations, columns, beams, girders, etc.), (2) walls and partitions (exterior and interior), (3) various types of floors and roof assemblies, (4) various floor and roof coverings.

For effective fire protection, the building construction components must be able to support normal structural loads during a fire, contain the spread of smoke and fire gases, and prevent excessive heat flow for a reasonable time period.

Framing

Structural framing members are usually composed of one or more of the following materials: reinforced concrete, pre-stressed concrete, steel, iron, aluminum, and wood.

The presence of noncombustible materials such as structural steel and iron in a building of otherwise masonry construction does not mean that the building is classified as fire resistive. If exposed iron and steel structural members are present, the entire building is open to danger from unequal rates of expansion of the masonry and metal members. Structural steel and iron will expand and distort at relatively low fire temperatures (at 1200° F, steel has loss of strength) unless properly protected.

Covering of structural members with a protective material is therefore necessary to insulate them against a rise in temperature which would impair or destroy their strength or usefulness by softening or expansion. This protection also prevents transmission of dangerous temperatures through walls and floors to other parts of the building.

It is necessary to have an understanding of the various types of materials used to produce fire-resistive construction. Before this, it should be noted that no construction material in existence can totally resist impairment from a fire. This is why the materials used to fire protect steel are termed fire resistive not fire proof.

Many of the types of materials used to produce fire-resistive construction are part of an encasement system. Gypsum is one such material. It delays the transfer of heat to steel by releasing water by crystallization which is evaporated. Gypsum is used in a plaster and can be applied or pre-cast onto steel

Portland cement delays heat transfer in the same manner as gypsum. It is used to encase steel and to decrease spalling when applied to concrete. Portland cement plaster has less effective fire resistance than gypsum plaster, but is used where direct exposure to humidity and the elements is unavoidable.

Perlite and vermiculite are lightweight aggregates with high thermal qualities. They are products of volcanic rock and are used as a substitute for sand in concrete or plaster. Plaster or cement with perlite or vermiculite are usually applied to the underside of ceilings or roofs.

Mineral fiber, as it used to be, contained concentrations of asbestos or free crystalline silica. Now it contains neither of these for health reasons. Mineral fiber is lightweight and is used as a spray coating on steel.

There are also coatings and treatments that can be applied to wood. One of these is a wood glue lamination. This process allows the wood to char partially but not burn all the way through.

Safety managers should be familiar with the types of materials used to design fire-resistive structural buildings so they can communicate with architects and engineers in discussions concerning fire-resistive design.

Walls and Partitions

There are four basic materials used for wall construction. These are: reinforced concrete, masonry, steel frame, and wooden frame. There are various types of wall and partition assemblies using different variations of these materials (Przetak, 1977).

Floor and Roof Assemblies

There are various types of floor and roof assemblies commonly used in the construction of buildings. Some floor assemblies and their supports are similar to that of roofs and their supports. The factors that determine the selection of floor and roof assemblies are:

1. the loads to be supported
2. durability and weight of the system
3. suitability for the installation of utilities
4. fire resistance requirements
5. temperature and sound insulating properties

The determination of the suitable fire resistance of a roof or floor is generally based on the fire resistance of the building as a whole.The fire resistance of a floor or roof may be inherent in the type of construction used, or it may be applied by the addition of fire resisting materials. A reinforced concrete slab on concrete beams represents a type of construction that has inherent fire resistance. A noncombustible roof deck or a reinforced concrete floor slab on unprotected steel has little fire resistance because the exposed steel would fail early under these conditions. Fire resistance may be provided for such assemblies by encasing the steel in fire resistance materials or protecting it with a suspended ceiling which would provide the desired fire resistance.

The reader may refer to the Underwriters Laboratories, Inc., and the Bureau of Standards for lists of the various types of floor and roof assemblies for their fire resistance in accordance with *NFPA 251*.

Roof and Floor Covering

There are many different types of roof coverings. These roof coverings range from highly flammable wood shingles to built-up coverings that are highly fire resistant. The particular hazards of each separate roof covering must be researched to see which best fits the need of a particular organization.

The Underwriters Laboratories, Inc., has set up three different classes of fire-retardant roof coverings. They are classified as A, B, and C. Class A coverings are the highest ranking possible. Roof coverings in this classification are effective against severe fires. Roof coverings of this class are not readily flammable and do not spread fire. They give a fairly high degree of fire protection for the roof, do not slip from position, possess no flying-brand hazard, and do not require very much maintenance to maintain fire retardance.

Class B coverings are roof coverings that provide medium fire protection. Class B coverings are not readily flammable and do not help the spread of fire. Also, like Class

A coverings, they will not slip from position and possess no flying-brand hazards, but, unlike Class A coverings, they require periodic repairs in order to maintain their fire-retardant properties.

Class C coverings are the lowest rating possible for roof coverings. These coverings are effective only in light fire exposures. Similar to Class A and B coverings, these coverings are not readily flammable, do not spread fire, do not slip from position, and possess no flying-brand hazard. But unlike Class A and B coverings this classification requires occasional repairs or renewals to maintain its fire-retardant properties. Most building codes require A or B coverings when these buildings are within the city limits and whenever fire-resistive buildings are specified in the building codes. In some codes Class C coverings are permissible. In many codes Class C coverings are listed as the minimum specified coverings allowable. It is important, when considering which covering to put on a building, to consult all appropriate building codes to see which coverings are permissible and which coverings are not. There are two main types that roof-covering classifications fall into: prepared and built-up coverings. Built-up coverings, as the name implies, consist of coverings that are *built up* layer by layer to form the roof covering. A built-up roof covering can come in many different fashions. A simple tar and gravel roof is a built-up roof covering. An example of the layers that might be applied in a built-up roof covering process might be:

- Layer 1 - wood deck
- Layer 2 - sheathing paper
- Layer 3 - layers of felt
- Layer 4 - mopping of asphalt
- Layer 5 - alternating layers of asphalt and felt
- Layer 6 - gravel or slag embedded in the pouring of asphalt

Prepared roof coverings consist of shingles or sheathing. These coverings can only be applied to roofs where nails can be driven. Prepared roof coverings must also have a severe enough incline so that water is not allowed to stand.

When combustibility of a roof is determined, the aspects that are looked at are the physical characteristics of the roof deck and its supports. Usually no consideration is given to the combustibility of the covering and insulation. These factors are too important to be overlooked. Insulation and coverings that are combustible do not usually affect the classification of the roof. In some large-scale fire testing of roofs, it has been proven that some metal roofs can contribute to the spread of fires.

One of the most important factors in the limitation of fire damage is the design of floors to resist the passage of heat, smoke, gases, and water from story to story. Fire usually spreads upward faster than downward; therefore, the underside of floor construction is more important than the upper surface. Thus, wooden flooring is commonly used over a concrete slab or other fire-resistive floor construction with little detrimental effect on the fire safety of the building. A hazard is present if it is so laid as to form a concealed space through which fire may spread under the finish floor. Floor surfacing of highly combustible types, which are easily ignited, ordinarily do not, of themselves, contribute to the spread of fire unless the combustible contents of the interior finish are involved in a fire which produces temperatures sufficient to cause a flashover.

Carpeting may contribute to fire spread depending on the length and density of the nap, the chemical structure of the material, and other factors. The test procedure now accepted by all jurisdictions for determining carpet acceptance is *NFPA 253*. The *Life Safety Code* (*NFPA 101*) uses this standard.

The tightness of wood floor surfacing is an important factor in buildings of combustible construction. Small sparks or cigarettes lodging in cracks between boards may cause ignition much more readily than when resting on top of a tight floor. Any cracks or holes are a source of obvious hazard.

Fire Loading

In order to determine the fire resistance needed for walls and floors, it is necessary to understand the quantity and characteristics of the combustibles stored within the structure, and of the structure's building materials. Brannigan (1982) offers this discussion of fire loading:

> The fire load is the potential fuel available to the fire. To the extent that the building is combustible, the building is part of the fire load, as are all combustible contents . . . (We) confine the term 'fire load' to simply the amount of fuel present and use the term "rate of heat release" (RHR) to describe the intensity of the fire, the rate at which the fuel burns.
>
> The basic measurement of the fire load is the British thermal unit (Btu), the amount of heat it takes to raise one pound of water one degree Fahrenheit (metric: 1 Btu = 1 kilojoule) . . .
>
> Each pound of combustible material has its own caloric value in terms of Btu per pound. For estimates, two rules of thumb are used. For wood, paper and similar materials, 8,000 Btu per pound is the caloric value; for plastics and combustible liquids, 16,000 Btu per pound is the caloric value.
>
> The weight of the fuel is multiplied by the appropriate Btu value and divided by the area to arrive at Btu per square foot, the measure of the fire load. You may find fire load expressed in pounds per square foot. . . i.e., wood and paper, estimated at 8,000 Btu per pound. . . . Plastics are converted into 'equivalent pounds' on the basis of 1 pound of plastic equals 2 pounds of wood.
>
> A fire load of 80,000 Btu per square foot or 10 pounds per square foot is taken to be the equivalent of the 1-hour exposure to the fire standard of the American Society for Testing and Materials E 119 Fire Endurance Test, 16,000 Btu per square foot is equivalent to 2 hours, etc.
>
> It is only recently (the 1970's) that the concept of calculating the fire load in advance has gained acceptance even in fire protection circles. Yet, the absorption of heat by water is the essence of fire suppression by fire departments.
>
> Many pre-fire plans note that a building is sprinklered. How many contain any analysis of whether the sprinkler system is capable of delivering the amount of water needed to absorb the heat that must be absorbed if the fire is to be stopped from spreading?
>
> It should be noted that the fire load is a measure of the maximum heat that would be released if all of the combustibles in a given area burned. Maximum heat release is the

product of the weight of each combustible multiplied by its heat of combustion. In a typical building, the fire load includes combustible contents, interior finish, floor finish, and structural elements. Fire load is commonly expressed in terms of the average fire load, which is the equivalent combustible weight divided by the fire area in square feet (m^2).

Equivalent combustible weight is defined as the weight of ordinary combustibles having a heat of combustion of 8,000 Btu per lb. (18,608 J/kg) that would release the same total heat as the combustibles in the space. For example, the equivalent weight of 10 lb. per sq. ft. (48.8 kg/m^2) of a plastic with a heat of combustion of 12,000 Btu per lb. (27,912 J/kg) would be:

10 lb. per sq. ft. x 12,000 Btu per lb. = 120,000 Btu per sq. ft.

120,000 Btu per sq. ft. ÷ 8,000 Btu per lb. ordinary combustibles = 15 lb. per sq. ft.

COMPARTMENTALIZING FACILITY

To prevent the spread of fire throughout the facility, it should be separated into compartments. Screens and curtains will do little to prevent fire spread, especially among easily ignited goods and objects. The most common, and most certain, fire prevention technique in compartmentalization is the use of a fire wall. This will serve to provide reasonable assurance that a fire will be contained to a specific area (Egan, 1978).

According to the NFPA, all fire walls and their supports shall be designed to withstand a minimum uniform load of 5 psf from either direction applied perpendicular to the face of the wall. In addition, all fire walls shall be nonload bearing. The structural framing within the plane of the wall shall be permitted to be load bearing. The best way to protect fire wall openings is to minimize the number of openings that exist within the structure. Fire walls are rated based on the total hours they are designed to withstand exposure to a fire, and the reason for this is that fire walls are designed to provide separation between the various building areas so that fire will not spread to other sections of the building.

Fire Doors and Windows

Fire doors are the most widely used and accepted means of protection of both vertical and horizontal openings in a building's structure. They are manufactured to specifications designed to produce a door capable of withstanding various degrees of fire exposure.

Fire door and fire window classifications are located in Appendix E of the *NFPA Fire Codes 2001*. The scope of the *NFPA 80* standard for fire doors and fire windows designates the following:

"This standard shall regulate the installation and maintenance of assemblies and devices used to protect openings in walls, floors and ceilings against the spread of fire and smoke within, into or out of buildings."

Fire doors, whether new or existing, are classified by the following designation system:

1. Hourly rating designation; or
2. Alphabetical letter designation; or
3. A combination of 1 and 2; or

4. If a horizontal access door, a special listing indicating the fire-rated floor, floor-ceiling, or roof-ceiling assembly for which the door may be used.

Fire windows are also classified by an hourly rating designation and fall under the *NFPA 257, Fire Testing for Windows and Glass Block*.

The following alphabetical letter designation was one method employed to classify the opening for which the fire door is considered suitable. Openings are classified as A, B, C, D, E, and F in accordance with the character and location of the wall in which they are situated. In each of the following classes, the minimum fire protection ratings are shown. Doors, shutters, or windows having higher ratings are acceptable. Class A openings are walls separating buildings or dividing a single building into fire areas. Doors for the protection of these openings have a fire protection rating of 3 hours. Class B openings are in enclosures at vertical communication through buildings (stairs, elevators, etc.). Doors for protection of these openings have a fire protection rating of 1 or 1.5 hours. Class C openings are in corridor and room partitions. Doors for protection of these openings have a fire protection rating of .75 hours. Class D openings are in exterior walls which are subject to severe fire exposure from outside of a building. Doors and shutters for the protection of these openings have a fire protection rating of 1.5 hours. Class E and F openings are in the exterior walls which are subject to moderate or light fire exposure, respectively, from outside of the building. Doors, shutters, or windows for the protection of these openings have a fire protection rating of .75 hours.

There are several types of fire door construction which have been approved, and generally follow terminology of the industry and of testing laboratory classification, and are offered for descriptive identification of available doors. Some of these are:

- **Composite Doors.** Composite doors consist of wood, steel, or plastic sheets bonded to and supported by a solid core material.
- **Hollow Metal Doors.** Hollow metal fire doors are of a flush or panel design with not less than 30-gauge steel faces. Flush door designs include steel stiffeners or honeycomb core material to support the faces. The voids between stiffeners may be filled with insulating material. Panel door designs are a stile-and-rail construction with insulated panels.
- **Metal Clad or Kalamein Doors.** Metal clad fire doors are swinging type only and are of flush or panel design consisting of metal-covered wood cores and stiles and rails and insulated panels covered with steel of 24 gauge or lighter.
- **Sheet Metal Doors.** Sheet metal fire doors are formed of 22-gauge or lighter steel and are of corrugated, flush steel, or panel design.
- **Rolling Steel Doors.** Rolling steel fire doors consist of steel or stainless steel interlocking slats to form a curtain of not less than 22 gauge, attached to an overhead barrel mounted on brackets for attachment to walls. The complete assembly includes the operating counterbalance enclosed in the barrel, automatic closing mechanism, the door guides, metal hood enclosure, and flame baffle.

- **Tinclad Doors.** Tinclad fire doors are of two- or three-ply wooden core construction, covered with 30-gauge galvanized steel or terne plate (maximum size 14 in. by 20 in.) (.36 m by .51 m) or 24-gauge galvanized steel sheets not more than 48 in. (1.22 m) wide. Face sheets shall be vented.
- **Curtain-Type Doors.** Curtain-type doors consist of interlocking steel blades or a continuous formed spring steel curtain installed in a steel frame.
- **Wood-Core-Type Doors.** Wood-core-type doors consist of wood, hardboard, or plastic face sheets bonded to a wood block or wood particle board core material with untreated wood edges.
- **Special Purpose Doors**, such as acoustical fire doors, and frame assemblies are available in single swings and pairs that are furnished complete with sound seals.
- **Single-Unit-Type Doors.**

When selecting windows, common glass or plate glass has no fire resistance classification, as the glass cracks and falls under the influence of even moderate heat. Wired glass windows in metal frames have a fairly high degree of resistance to fire, but their use is restricted by the inherent limitations of the glass, which transmits radiant heat and flows at temperatures often reached in fires. Wired glass windows are recommended for use in moderate or light exposure. The period of resistance can be greatly enhanced when used in combination with a sprinkler system.

Fire Protection Rating

"The National Fire Protection Association does not approve, inspect, or certify any installations, procedures, equipment, or materials nor does it approve to evaluate testing laboratories." (NFPA, 1995)

The National Fire Protection Association's *NFPA 80, Fire Doors and Fire Windows*, standard is the determination (in minutes or hours) that fire doors, shutters, or windows have withstood a fire exposure as established in accordance with the procedures of *NFPA 252, Fire Tests of Door Assemblies*, which could be considered a fire protection rating category.

Smoke and Heat Venting

There has been a general trend, since the end of World War II, toward large single-area, one-story buildings using light construction, to gain an increased efficiency in assembly line operations. Production-minded industrialists do not favor division walls as they restrict the mobility of conveyor lines and make operational changes and expansions more difficult. The result, from the fire protection viewpoint, has been the very vulnerable exposure of high values, within large, single fire areas, to extensive fire loss with accompanying production interruption of staggering proportion.

This has increased the difficulty of fire fighting, since the fire department must enter these buildings to combat fire in the central sections of the plant. If unable to enter the building, because of heat and smoke, efforts may be reduced to vain application of hose streams to the perimeter areas, while fire consumes the vast interior of the plant.

Fire extinguishment is normally accomplished by absorption of heat by water applied by sprinklers or hose streams with resultant reduction of the temperature of the burning material below its ignition point. The release of heat from its confinement within a building, through proper venting equipment, reduces the amount of required cooling and generally retards spread of the fire (*Fire Protection Handbook*, 1981).

Vents are not a substitute for sprinklers or other extinguishing facilities. Their purpose is to relieve smoke and heat from the building and to improve accessibility for the fire department so as to permit close approach and direct action against the seat of the fire.

Application and Scope

These provisions are intended to offer guidance in the design of facilities for the emergency venting of heat and smoke from uncontrolled fires. They do not attempt to specify under what conditions venting must be provided as this is dependent upon an analysis of the individual situation. However, venting is particularly desirable in those situations where manual fire fighting may be unduly handicapped or where automatic protection may be overtaxed as, for example, in large-area industrial buildings or warehouses, windowless buildings, underground structures, or in areas housing hazardous operations.

This guide does not apply to other ventilation (or lighting, as may be the case with monitors and skylights) designed for regulation of temperatures within a building, for personnel comfort, or production equipment cooling.

Venting may be desirable in either sprinklered or unsprinklered buildings. A serious fire may occur during a period when all, or a portion, of the automatic sprinkler protection may be out of service for repair or changes. In addition, a fire, in concentrated operations involving highly combustible materials and warehousing, may spread rapidly and overtax the sprinklers. Because of this, the combined counteracting effect of heat and smoke relief and fire department action may be essential to check its spread. Building construction of all types is included, although it is to be recognized that superior fire-resistive construction has inherent advantages.

Principles of Venting

There are so many variables which apply to the burning of combustible material that no exact mathematical formula is possible for determining precise venting requirements. The rate of combustion varies appreciably according to the nature, shape, size, and packaging of the combustible material, the size and height of piling, and other factors; the volume of heat and smoke to be vented differs accordingly. Vent sizes and ratios have therefore been developed from tests and experience, using theory only for guidance. If severe damage to exposed structural steel is to be avoided, temperature of vented heat must not be sufficient to overheat the steel, thus materially reducing its strength (Whitman, 1979).

The height of a column of hot gases has a direct relationship to the volume of hot gas that will be discharged by thermal updraft through an opening of a given size. Curtain boards, or their equivalent, increase the column effect which is essential to good venting.

Classification of Occupancies

Tests and studies provide a basis for division of plants into classes depending upon the fuel available for contribution to a fire. There is a wide variation in the quantities of combustible materials in the many kinds of industrial plants, and between various buildings and areas of most any individual plant. Classification should take into account the average or anticipated fuel loading and the rate of heat release anticipated from the combustible materials or flammable liquids contained therein.

1. Low Heat Release Occupancies: This class includes those buildings or portions of buildings containing scattered, small quantities of combustible materials. Such areas might be found in:

    Metal stamping plants
    Machine shops, with dry machining and like operations
    Foundries
    Breweries
    Dairy products processing plants
    Bakeries
    Meat packing plants

2. Moderate Heat Release Occupancies: This class includes those buildings or portions of buildings containing moderate quantities of combustible material which are fairly uniformly distributed. Such areas might be found in:

    Automobile assembly plants
    Leather goods manufacturing
    Printing and publishing plants
    Machine shops using combustible oil coolant, hydraulic fluids, or involving similar hazards

3. High Heat Release Occupancies: This class includes buildings or portions of buildings containing either hazardous operations or concentrated quantities of combustible materials or both. Such areas might be found in:

    Painting departments
    Oil quenching departments
    Chemical plants
    Paper mills
    Rubber products manufacturing plants
    General warehouses

It is to be recognized that many plants will have buildings or areas falling into each of the above classifications. An automobile plant, for example, might contain stamping presses and dry machining (low heat release); upholstery and trim (moderate heat release); and large paint spraying and dipping operations, and rubber tire storage (high heat release). Accordingly, venting facilities should be designed for the different classifications.

Vents

The following are types of vents.

1. **Monitors.** This type usually depends upon the breakage of ordinary glass (not over 1/8-inch thick) in the side walls to provide venting, although, where light is unimportant, metal panels may be used in lieu of glass, and arranged to open automatically in event of fire. Where conservation of building heat is not a factor, louvers are often used. Wired glass is unacceptable unless the sash is arranged to open automatically. Both sides of a monitor should be designed to vent to assure that wind direction, at time of a fire, will not impede its effectiveness.
2. **Continuous Gravity Vents.** This type of vent is a continuous, narrow slot opening, with a weather hood above, similar to those frequently used along the gable or a pitched-roof, foundry-type building. If movable shutters are provided to control temperature, they should be automatic-opening in the event of a fire.
3. **Unit-Type Vents.** This type of vent is of relatively small area, usually 4 by 4 feet to 10 by 10 feet, and is distributed about the roof according to the occupancy requirement. Generally they are lightweight metal frames and housing, with hinged dampers which may be operated manually or automatically opened in event of fire.
4. **Sawtooth Roof Skylights.** Since wired glass in a fixed sash is generally used in sawtooth skylights, it offers no value as a venting facility, unless a plain glass is used or it is a movable sash equipped with devices to open automatically in case of fire.
5. **Exterior Wall Windows.** These may be considered as effective vents provided the windows are along the eaves. Lower windows are of very limited venting benefit since heat will bank up against the ceilings. In multiple-story buildings, exterior windows may be the only practical means of venting of all but the top floor.

Release Methods

It is essential that release of the venting facility be automatic in operation to eliminate the uncertainty of the human element. The release should be relatively simple in design and independent of electrical power since electrical services may be interrupted by the fire.

Automatic operation is best secured by a simple linkage with a fusible link in connection with counterweights, and associated equipment, utilizing the force of gravity for opening the vents. It is permissible to utilize the vents of normal ventilation by means of motor-driven or manually operated shutters, dampers, covers, and like equipment.

However, an automatic release is still essential and must be capable of releasing the vent independently of any other device.

Release devices which permit automatic opening from internal pressure are undesirable over occupancies which are susceptible to water damage. Vents so equipped may open as a result of pressure differential during wind and rain storms. Authorities having jurisdiction should be consulted.

Venting Ratios

The following ratios of effective area of vent openings to floor areas should be provided for the various occupancy classifications:

1. Low Heat Release Content: 1:150
2. Moderate Heat Release Content: 1:100
3. High heat Release Content: 1:30 to 1:50

Explosion Hazards

In any proposed use of a suppression system where there is a possibility that personnel will be in locations receiving a suppressant discharge, suitable safeguards shall be provided to insure prompt evacuation. These may include, but are not limited to, personnel training, warning signs, discharge alarms, and the use of breathing apparatus.

Determination of Deflagration Characteristics

The design of a suppression system is based upon the (a) rate of pressure rise, and (b) rate of burning, etc., of a deflagration during its incipient stages. Extensive testing has been accomplished on a wide range of materials, but as new materials and processes are developed, tests shall be performed on them.

Determination of Hazard to be Protected

A thorough hazard analysis shall be performed to establish the type and degree of explosion hazard inherent in the process. Such factors as the type and ratio of combustible and oxidant, total volume to be protected, critical operating conditions, etc., shall be reviewed, as well as possible malfunction situations that may affect the extent of the hazard. It is vital that this analysis be as extensive and complete as possible to assure that maximum protection is provided by the suppression system.

EXPLOSION PREVENTION SYSTEMS

Detecting an incipient deflagration can be done by sensing either the pressure increase or radiant energy from the combustion. Temperature sensors are normally too slow for use in a suppression system. Under unusual circumstances, however, this method of detection may be applicable, but tests shall be performed to confirm the effectiveness.

Detectors which respond to a rate of pressure rise from the incipient deflagration are applicable where pressure fluctuations are encountered in the process. Caution shall be exercised using this sensing mode to assure that sufficiently rapid detection is achieved over the entire flammable range.

Static pressure detectors are used where a constant pressure level is normal. A fast detection response is essential, along with maximum pressure sensitivity. Radiation sensors are used as detection devices for suppression systems under special circumstances. The complexity and extreme sensitivity of such devices require a complete engineering analysis to assure proper operation.

EXPLOSION SUPPRESSION SYSTEMS

Explosion suppression is a technique by which burning of a confined mixture is detected and arrested during its incipient stage, preventing development of pressure which could result in explosion.

The suppression system shall be capable of sensing the predominant characteristic of the incipient deflagration. Careful consideration is necessary when designing systems for facilities where chemical or other processes may induce a wide variation in pressure and temperature conditions to assure that the detection device functions correctly over the entire range of typical conditions. In addition, the selection of the suppressant shall be made with consideration of the possible chemical reaction between it and the materials that may be encountered. Proof of the compatibility of the suppressant shall be established.

After the type of equipment that will comprise a system has been established, the proper location of detectors and suppressers shall be determined. The necessity for rapid detection and high speed dispersion of the suppressant requires the careful study for proper installation of the suppression equipment.

Explosion suppression systems can be used where combustible gases, mists, or dusts are present within enclosures and where the suppressant can be effectively distributed. The following may be protected by explosion suppression systems:

- Processing equipment, such as enclosed reactors, mixers, blenders, pulverizers, mills, dryers, ovens, filters, screens, dust collectors, etc.
- Storage equipment, such as atmospheric or low-pressure tanks, pressure tanks, mobile facilities, etc.
- Material-handling equipment, such as pneumatic and screen conveyors, bucket elevators, etc.
- Laboratory and pilot plant equipment, such as hoods, glove boxes, test cells, and equipment indicated above.

There are limitations in the use of explosion suppression systems which should be recognized. Such limitations involve the nature, size, and geometry of the equipment to be protected and the physical and chemical properties of the reactants. Applications of suppression systems require careful study because of the technical complexity of the hazard.

Explosion Venting

Vents do not prevent the occurrence of a deflagration, but are intended to limit the damage from the pressure generated by the deflagration. This guide applies to the deflagration of combustible dusts, gases, or mists when mixed with air during manufacturing operations and storage. Typical examples of industrial equipment to which this applies include crushers, grinders, pulverizers, sieves, screens, bolters, dust collectors and arrestors, conveyors, screw feed conveyors, bucket elevators, dryers, ovens and furnaces, spray dryers, blenders, mixers, ducts, pipes, bins, silos, spreaders, coating machines, and packaging equipment.

Venting Deflagrations

A vent in an enclosure (building, room, or vessel) is an opening through which newly formed or expanding gases may flow. The purpose of the vent is to limit the maximum pressure from a deflagration in order to limit damage to the enclosure. Extensive destruction may result if combustion occurs within an enclosure too weak to withstand the full force of the deflagration. An ordinary building wall (8-inch brick or an 8-inch concrete block) will not withstand a sustained internal pressure as small as 1 psig (144 lbs. per sq. ft. or 6.9 ka).

Unless the enclosure is designed to withstand the maximum pressure resulting from a possible deflagration, venting should be considered to minimize damage due to rupture. The area or the vent opening must be sufficient to limit pressure build-up to a safe value. Combustion venting of an enclosure normally implies the need to vent in such a manner that the maximum pressure development is low. The maximum pressure should be lower than the pressure which the weakest building or structural member can withstand. The weakest building member may be a wall, roof, or floor of the enclosure if it is elevated. On equipment the weakest section may be a joint.

Description of Vents and Vent Closures

The vents described in this section have been designed or developed for the release of pressure effective only in deflagrations in which explosions of dusts or gases may occur. In most cases, the described vents are effective only in deflagrations in which the rate of pressure rise is moderate and where, in large enclosures, only a part is involved in the deflagration. The devices described are not generally suitable for protection of pressure vessels, which is outside the scope of this guide, nor for protection against pressure.

Some types of vent closures are commercially available and may be purchased ready to install in buildings or equipment. The following descriptions should be used as the basis for development of suitable vents and vent closures which will provide the desired protection.

Open or unobstructed vents are the most effective vents for the release of deflagration pressure from enclosures. They provide an unobstructed opening. However, there are comparatively few operations with inherent deflagration hazards that can be conducted in open equipment installed in buildings without walls. Often some form of vent closure must be provided to protect against the weather, to conserve heat, to bar unauthorized entry, to preclude dissemination of the combustible material, or to prevent contamination of the product by the entrance of dirt or moisture from the outside. Open equipment is recommended wherever a more serious deflagration hazard is not created through dispersion or dissemination of the material and where closed equipment is not necessary to prevent contamination of the material.

Louvers. Although openings containing louvers cannot be considered completely unobstructed vents, they do provide a large percentage of free space for the release of deflagration pressure and have served effectively as vents. They are recommended especially as wall vents where windows are not required to maintain controlled atmosphere conditions within the enclosures. Louvers can be used effectively as vents where it is necessary or desirable to prevent unauthorized entry or egress. However, compensation for pressure drop must be considered.

Hanger-Type Doors. Large hanger-type or steel curtain doors installed in side walls of rooms or buildings can be opened to provide unobstructed vents during the operation of any process or equipment in which there is an inherent deflagration hazard. Such doors can be closed to prevent unauthorized entry when the equipment is unattended or not in operation. This type of venting has been effective and is highly recommended, but strict supervisory control is essential in cold climates to insure that employees do not sacrifice safety for comfort by keeping the doors closed during operation.

Open Roof Vents. Large roof openings protected by weather hoods can serve as deflagration vents on one-story buildings or the top story of a multiple-story building. This type of venting is effective particularly where lighter-than-air gases may escape from processing equipment and create a hazard near the ceiling of the enclosure. In addition to serving as vents for the release of pressures, such roof openings reduce the possibility of a deflagration by providing a channel through which the gas can escape from the building.

Closed or Sealed Vents. Where large openings cannot be permitted in a building, the most desirable arrangement is an isolated single-story building. Such a building can be most easily designed for explosion resistance and venting. Equipment which requires venting should be located close to outside walls so that ductwork, if necessary, can be short.

Building vent closures are necessary in air-conditioned plants or where heat is provided for the comfort of occupants during all or part of the year. Vent closures are required on processing equipment whenever it is necessary to retain dust or gas or where processes are conducted under pressure, vacuum, or other controlled atmospheric conditions. The fundamental principle in the design of vent closures is that the vent will open at as low a pressure as possible. It should have no counterweights; counterweights add to inertia.

INSTALLATION OF UTILITIES AND SERVICES

Most building codes define building utility systems as those systems that are essential to the functioning of a building for its designed purpose and that normally remain with the building if the owner or tenant moves out with or without his machinery or other equipment. Building utilities include all parts of the utilities up to and including the tie-ins to the corresponding utility systems that have been or will be installed to serve particular equipment or operational needs. Building utility systems do not, in general, include any part of the auxiliary systems.

All building utility systems have one fire safety requirement in common. Whenever the utility design requires piercing a necessary fire barrier (walls, floors, roofs), the design should be such that the essential integrity of the fire barrier is maintained. You should be careful to see that poke-through assemblies do not breach a fire barrier.

Plant utilities and services that are to be discussed include the following:

1. Electrical Installation
2. Gas Piping Installation
3. Elevators, Dumbwaiters, and Vertical Conveyers
4. Rubbish Chutes, Incinerators, and Laundry Chutes
5. Electronic Computers and Data Processing Equipment

Electrical Installation

With all electrical systems, be certain that the proper types of protection equipment have been installed and are being maintained and that adequate fire protection is present. Electrical wiring and equipment installed should be in accordance with the *NFPA 70, National Electrical Code*.

Gas Piping Installations

Modern gas systems serve many types of uses, including central heating, processing ovens, chemical laboratories, and unit heaters. The pressure in a system may vary from low to high, depending on operating requirements. The variations in pressure may be great enough to require the provision of special control equipment.

Building gas systems are valuable tools in any industrial society if installed and used properly and maintained with due regard to the explosive and fire potential. Misused, they can be the source of major disasters that could have been avoided by proper attention to a few fundamentals.

Equipment utilizing gas and related gas piping should be installed in accordance with *NFPA 54, National Fuel Gas Code*, or *NFPA 58, Liquefied Petroleum Gas Code*. Existing installations may be continued in service, subject to approval by the authority having jurisdiction. Review all national fire codes published by the NFPA that pertain to this subject matter, as there are several standards that may be applicable.

Elevators, Dumbwaiters, and Vertical Conveyers

Elevators, passenger and freight, including the power supplies thereto, are almost always considered 100-percent building utilities. Exceptions to this rule are few. Elevators located in mechanically pressurized vestibules or smoke-proof towers can be used by fire fighters for rescue and suppression operations and by elderly and handicapped persons for limited emergency escape or movement to a refuge area. Elevators are normally designed to move a certain percentage of the building population during the peak 5 minutes of elevator demand, not to move everyone when they try to leave at the same time. Under these design conditions, complete evacuation of buildings by elevators would take 20 to 45 minutes for offices, $1\frac{1}{4}$ to $1\frac{3}{4}$ hours for apartments, and $\frac{3}{4}$ to $1\frac{1}{4}$ hours for hotels.

Elevators shall not be considered an exit component. [Exception: when specifically permitted by individual occupancy (Chapters 7 through 16 of the *Life Safety Code*, *NFPA 101*, Chapter 7: Building Service and Fire Protection Equipment: Section 7–4, Elevators, Escalators and Conveyors), and when measures satisfactory to the authority having jurisdiction are taken to provide smoke control for the elevator shaft.] Elevators shall be installed in accordance with the *Safety Code for Elevators, Dumbwaiters, Escalators, and Moving Walks* (ANSI A17).

Vertical conveyers, including dumbwaiters and pneumatic conveyers serving various stores in a building, should be separately enclosed within walls or partitions in accordance with the provisions of Section 6–1, *NFPA Life Safety Code 101*, Chapter 6: Features of Fire Protection. Service openings should not open to an exit. Service openings, when required to be open on several stories at the same time for purposes of operation of the conveyer, should be provided with closing devices which will close all service doors upon activation of smoke detectors, which are located inside and outside the shaft enclosure in locations acceptable to the authority having jurisdiction.

Rubbish Chutes, Incinerators, and Laundry Chutes

Each rubbish chute or chute to an incinerator should be separately enclosed within walls or partitions in accordance with the provisions of Section 6–1, *NFPA Life Safety Code 101*. Openings serving chutes and incinerator flues should be protected in accordance with Section 6–1. Doors for such chutes or incinerator flues should not open directly to an exit, corridor, or normally occupied room, but should open to a separated room or closet. The room or closet should be separated from other spaces in accordance with Sections 6–5 and 6–6 of the *Life Safety Code*.

Laundry chutes should be enclosed and protected in the same manner as rubbish chutes. Rubbish chutes, laundry chutes, and incinerators should be installed and maintained in accordance with *NFPA 82, Incinerators and Handling Systems for Waste and Linen.* Existing installations may continue in service, subject to approval of the authority having jurisdiction.

Electronic Computers/Data Processing Equipment

With all business operations now depending much more on computers and electronic equipment that is very expensive to purchase, this equipment should be installed in specifically designated areas. Computer and data processing equipment may incur damage when exposed to elevated, sustained ambient temperatures. The following summarizes expected damage to computer equipment exposed to elevated temperatures:

1. Damage to computer equipment may begin at a sustained ambient temperature of 175° F (79.4° C), with the degree of damage increasing with further elevations of the ambient temperature and exposure time.
2. Damage to magnetic tapes, computer disks, and other computer storage media, may begin at sustained ambient temperatures above 100° F (37.8° C). For this reason, such computer storage media are normally stored in a separate room next to the computer room. This falls under *NFPA 75, Standard for the Protection of Electronic Computer/Data Processing Equipment.* Damage occurring between 100° F and 120° F (48.9° C) can generally be reconditioned successfully, but the chance of successful reconditioning lessens rapidly with temperatures higher than 120° F.
3. Damage to disks may begin at sustained ambient temperatures above 150° F (65.6° C) with the degree of damage increasing rapidly with further elevations of sustained ambient temperatures.
4. Damage to paper products may begin at a sustained ambient temperature of 259.1° F (176.6° C). Paper products which have not become brittle will generally be salvageable.
5. Damage to microfilm may begin at a sustained ambient temperature of 225° F (107.2° C) in the presence of steam or at 300° F (154.4° C) in the absence of steam.

Because computers and other electronic equipment are susceptible to fire damage and the heat, steam, and combustion that follows, fire protection of this equipment is of critical importance. Once management commits itself to a program of dependency on any such equipment, simple economics dictates doing away with former methods and procedures. The personnel, equipment, and facilities are no longer available to pick up the load assumed by the data processing equipment if it is put out of operation by fire or other unforeseen occurrences. Often, the major loss caused by disruption of the computer operation is from business interruption rather than from the actual monetary loss represented by the equipment itself, although the latter may run into millions of dollars.

Electronic computer/data processing equipment should be installed and maintained in accordance with *NFPA 75, Standard for the Protection of Electronic Computer/Data Processing Equipment.*

HEATING, VENTILATING, AND AIR CONDITIONING

Heating appliances are causes of concern for the safety professional because of the temperature at which they operate—above the ignition temperature of many materials. Also, there is always the hazard of unburned fuel discharging or the explosion of the fuel. The best approach is to keep combustible materials excluded from the spaces to be protected. There are three methods of minimizing the risks associated with combustibles: (a) containment, (b) ventilation, and (c) purging. Obviously, it is important to have a baisc understanding of the fuel the business uses when one is establishing a fire prevention plan. (More information may be found in the *Fire Protection Handbook*, 18th Edition, Explosion Prevention and Protection, Section 4/Chatper 14.)

Fuels are either solid, liquid, or gas. Some solid fuels are coal, wood pellets, charcoal, briquettes, peat, bagasse, and sugar cane. Liquid fuels would include fuel oil or alcohol. As each fuel has its own method of firing, the wise safety professional would need to investigate the properties of the fuel being used. Additionally, he or she must study how that fuel is stored and then transported to the heating appliance.

Explosions within fireboxes are not common, but if they do occur extreme damage can follow. Therefore, one must protect against the uncontrolled flow of unburned fuel into appliances. This may be done through the use of various types of controls, which are often mandated by law. Some types of controls are:

- Primary safety controls
- Air fuel interlock
- Pressure regulation and interlock
- Oil temperature interlock
- Manual restart
- Remote shutoff
- Safety shutoff valves
- Safety control circuits

As each appliance is different, so is each control. They all have their positive and negative aspects in terms of fire potential. Proper checks and maintenance of these pieces of equipment would be the foremost precaution in existing equipment. Proper selection of such in a newly constructed facility would be, of course, most desirable. When discussing air conditioning and ventilating, one should refer to *NFPA 90A, Installation of Air Conditioning and Ventilating Systems*, for details; however, a brief look at the topic follows. All systems of this sort should be designed to:

- avoid any combustibles within the duct system, including filters, duct liners, and the duct construction itself.

- avoid combustible exterior duct insulation.
- prevent the passage of smoke and fire through the ducts.
- maintain the integrity of fire and fire-barrier walls where penetrated by ducts.
- maintain the integrity of floors where penetrated by ducts or connectors.

There are several different types of air conditioners, all which supply or remove air naturally or mechanically. The main system (i.e., fans, heaters, filters, etc.) is located in an area cut off from the rest of the facility by walls, floors, and ceilings that provide a minimum fire resistance rating of 1 hour. Such an arrangement will keep a fire from spreading to other areas for the specific rating period.

Fresh air intakes of a system must be strategically located, as undesirable outside air could be drawn in and spread throughout the building. If that is the case, dampers that can be controlled by fire and smoke detectors should be installed at the intakes. As smoke generally rises, intakes installed low lessen the possibility of drawing in smoke.

Air cooling and heating equipment poses two basic hazards: electrical and refrigerant. Therefore, proper installation is a must. For further information refer to the *National Electrical Code* and the *Safety Code for Mechanical Refrigeration*.

There are three types of air filters and cleaners used in air conditioning and ventilating systems: electronic air cleaning, renewable media, and fibrous media units. All remove particulates from the air and if not cleaned may ignite and spread smoke throughout the entire system. For this reason, smoke sensors should be installed and proper maintenance must be a top priority.

In areas where smoke control or exhaust systems are required, they should conform to the requirements of the building code authority having jurisdiction (*NFPA 92A, Smoke-Control Systems*).

Ducts to any ventilating system can become a means of distributing smoke if precautions are not utilized. As ducts probably will pass through a wall, floor, or ceiling, regard for fire-stopping must be considered. Dampers offer an effective method of controlling this potential hazard. *NFPA 90A, Installation of Air Conditioning and Ventilating Systems (70:440)* provides whatever information one may need in order to protect buildings in this way.

MAINTENANCE

Repair service includes periodically recurrent inspections and tests required to keep the heating, ventilating, and air conditioning system and its component parts fully operational at all times, together with replacement of the system or its components, when for any reason they become undependable or inoperable.

The key to safely operating air conditioning and ventilating systems is the implementation of a maintenance program. Check the condition of the filters and electrical wiring and examine air ducts for dust and lint. Additionally, while inspecting the equipment, the maintenance crew should look for signs of rust and corrosion, especially on moving parts (*NFPA 90A, Installation of Air Conditioning and Ventilating Systems*).

SUMMARY

In order to meet the standards presented here, safety professionals must be able to utilize various resources. It is imperative they have ready access to the NFPA Codes mentioned in this chapter and have a working relationship with various community resource personnel. The state fire marshal, local fire department, city building inspector, and local building contractor each have their own area of expertise. Utilizing these resources can enable the practitioner to garner invaluable knowledge on the subject.

The key to fire control is, of course, prevention. Prevention can be highly utilized in any construction. Here the safety professional benefits from the fact that the systems discussed in this chapter are required by law to fulfill certain criteria. Thus armed, the safety professional can formulate programs which will prevent fire (Whitman, 1979). After a facility has been erected, careful maintenance combined with an effective fire prevention plan is imperative.

ADDRESSES

Building Officials and Code Administrators International Inc.
4051 W. Flossmoor Rd.
Country Club Hills, IL 60478-5795
(708) 799-2300

International Conference of Building Officials
5360 S. Workman Mill Rd.
Whittier, CA 90601
(213) 699-0541

Southern Building Code Congress International
900 Montclair Rd.
Birmingham, Al 35213
(203) 591-1853

Council of American Building Officials
5203 Leesburg Pike, Suite 708
Falls Church, VA 22041
(703) 931-4533

National Conference of States on Building Codes and Standards
505 Huntmar Park Rd., Suite 210
Herndon, VA 22070
(703) 437-0100

World Organization of Building Officials
Site 18 Box 31 SS1
Calgary, Alta. T2M 4N3
(403) 268-3441

REFERENCES

American Society for Testing and Materials. *ASTM Standards*. Philadelphia: American Society for Testing and Materials (issued annually).

Approval Guide. Norwood, MA: Factory Mutual Engineering Corporation (issued annually).

Blake, Roland P. *Industrial Safety*. 3d ed. Englewood Cliffs, NJ: Prentice-Hall, 1972.

Brannigan, Francis L. *Building Construction for the Fire Service*. 2d ed. Quincy, MA: National Fire Protection Association, 1982.

Building Materials Directory. Northbrook, IL: Underwriters Laboratories, Inc.

Building Officials and Code Administrators, Inc. *BOCA National Building Codes*. Country Club Hills, IL: Building Officials and Code Administrators, Inc., 1995.

Carson, P. E., and Wayne, G. "Who Sets Means of Egress Standards?" *NFPA Journal* 87 (5) (1993):45–48.

Cavanaugh, Casey Grant. "Triangle Fire Stirs Outrage and Reform." *NFPA Journal* 87 (3) (1993):73–82.

Cote, Arthur, and Bugbee, Percy. *Principles of Fire Protection*. Quincy, MA: National Fire Protection Association, 1988.

Cummings, P. E., et al. "ADA Sets a New Standard for Accessibility." *NFPA Journal* 87 (3) (1993):42–47, 92–96.

Egan, David. *Concepts In Building Fire Safety*. New York: John Wiley & Sons, 1978.

Fire Resistance Directory. Northbrook, IL: Underwriters Laboratories, Inc.

Klem, Thomas J. "25 Die in Food Plant Fire." *NFPA Journal* 86 (1) (1992):29–35.

Lathrop, James K. "Life Safety Code Key to Industrial Fire Safety." *NFPA Journal* 88 (4) (1994):36–46.

Lathrop, James K., and Birk, David, P. E. "Building Life Safety With Codes." *NFPA Journal* 86 (3) (1992):42–52.

Marchant, Eric W. *A Complete Guide to Fire and Buildings*. Great Britain: Harper & Row Publishing Co.

National Fire Protection Association. *Fire Protection Handbook*. Quincy, MA: National Fire Protection Association, 1995.

______. *Fire Safety Code, Standard 495, Explosive Materials Code*. Quincy, MA: National Fire Protection Association, 1995.

______. *Fire Safety Code, Standard 257, Fire Testing for Windows & Glass Block*. Quincy, MA: National Fire Protection Association, 1995.

______. *Fire Safety Code, Standard 252, Fire Tests of Door Assemblies*. Quincy, MA: National Fire Protection Association, 1995.

______. *Fire Safety Code, Standard 70, National Electric Code*. Quincy, MA: National Fire Protection Association, 1995.

______. *Inspection Manual*. 7th ed. Quincy, MA: National Fire Protection Association, 1994.

______. *Life Safety Code Handbook*. Quincy, MA: National Fire Protection Association, 1995.

______. *National Fire Codes*. Quincy, MA: National Fire Protection Association (published annually).

______. *NFPA 101: Code for Life Safety from Fire in Buildings and Structures*. Quincy, MA: National Fire Protection Association, 1995.

______. *NFPA 101, Life Safety Code*. Quincy, MA: National Fire Protection Association, 1994.

National Safety Council. *Accident Prevention Manual For Industrial Operations*. 10th ed. Chicago: National Safety Council, 1995.

Planer, Robert G. *Fire Loss Control*. New York: Marcel Dekker, 1979.

Przetak, Louis. *Standard Details for Fire-Resistive Building Construction*. New York: McGraw-Hill Book Company, 1977.

Smith, E. E., and Harmathy, T. Z. *Designing Buildings for Fire Safety*. Philadelphia: American Society for Testing and Materials.

Whitman, L. *Fire Prevention*. Chicago: Nelson-Hall, 1979.

STUDY GUIDE QUESTIONS

1. What factors should be considered in plant layout and why?
2. What factors should be considered in determining a plant location?
3. Where would the safety professional find pertinent information on construction materials and why is this information necessary?
4. What are the three different classifications of fire retardant roof coverings according to Underwriters Laboratories, Inc.?
5. What are the three classifications of occupancies under which industries are categorized according to fuel available for contribution to fire?
6. Plant utilities and services include computers and data processing equipment. When thinking in terms of a fire prevention plan, what special features should be included for this equipment? Why?
7. Floor design is an important factor in planning fire safety, why?
8. What are the classifications for openings when considering fire doors and windows?
9. What is the one fire safety requirement that all building utility systems have in common? Explain.
10. At what point in building construction should fire safety be implemented? What follow-up actions should be taken?
11. What is fire loading? How does one determine fire loading?

CASE STUDIES

1. You are the safety professional for a manufacturing firm. The firm has purchased an older plant. The management wants you to make the facility fire safe. What problems might you encounter? Develop a plan of action for the facility.
2. Your company is considering expanding into a new location. This will require the building of a new facility. Develop recommendations for construction of the new facility.

Fire Detection Systems

CHAPTER 6

CHAPTER CONTENTS

FIGURES

Automatic fire detection systems allow the presence of fires to be detected quickly. Quick fire detection is paramount to life safety and property conservation. Several types of fire detection systems are available on the commercial market. Each type of system has its own advantages and disadvantages. This chapter reviews the basic types of fire detection systems.

AUTOMATIC FIRE DETECTION SYSTEMS

A fire detection system is a system composed of numerous components and assemblies. Its purpose is twofold: detect the presence of a fire and initiate a warning to building occupants. The presence of a fire is detected by sensing the byproducts of combustion or heat production. Once a fire detection system senses the presence of a fire, it transmits an audio and visual alarm (bells and flashing lights) to the building occupants. Upon receiving the alarm, the occupants have time to evacuate from the building. Some fire detection systems also transmit a signal that can summon the fire department.

The most basic fire detection system is a person in good mental and physical health. People can detect the presence of a fire via their senses and warn others of the impending danger. People are, however, unreliable fire detectors because they are not always present in a building when a fire starts. Even when they are present, people may be unable to comprehend and evaluate the first signs of a fire.

A more reliable fire detection system is an automatic system. Automatic fire detection systems respond by transmitting signals through a pneumatic, electric, hydraulic, or mechanical communications system. Fire detectors are designed to respond to a fire when the physical-chemical conditions exceed predetermined response thresholds. Fire detection systems initiate an audible-visual signal to alert the building occupants and/or the local fire department when the physical-chemical condition exceeds the predetermined response threshold.

Problems may occur with automatic fire detection systems. These systems do not have the capability to determine the cause of a fire nor a fire's intensity. False alarms can also become a problem due to a system's inability to evaluate conditions. False alarms are not normally caused by a shortcoming of the system itself. Improper system selection, random placement of detectors, poor installation, and human error are more common causes of false alarms.

Automatic fire detection systems are designed to respond to one or all of the three major physical-chemical processes involved in the conversion of energy and matter during a fire. During a fire, a particular physical-chemical environment is created. This process produces three major byproducts: thermal energy, radiation, and airborne particulates. The three classifications of fire detectors adopted in *NFPA 72, National Fire Alarm Code*, are based on heat, flame, and smoke detection (Grabowski, 1972). The thermal energy produced by a fire causes both a laminar and turbulent air flow. This heated air movement is detected by a thermal detector designed to activate when the temperature of the air has reached a predetermined threshold. Radiation is also produced by a fire in the ultraviolet spectrum below 4000 Angstrom, in the visible between 4000–7000 Angstrom, and in the infrared spectrum above 7000 Angstrom. The radiation given off by a particular fire depends on the intensity of the fire and the type of materials being burned. Flame detectors are designed to detect both ultraviolet and infrared radiation (Grabowski, 1972). The third byproduct of a fire is airborne particulates. These particulates are aerosols and usually range in size from 0.01 to 10 microns. Airborne or smoke particles smaller than 5 microns are not visible to the human eye. Smoke detectors respond to both visible and invisible particles. They are classified as either ionizing type or photoelectric type by Underwriters Laboratories.

Automatic fire detection systems can function not only as alarm systems for building occupants, but they can perform many other functions. Fire detection systems can function as initiating releasing devices for extinguishing systems. Extinguishing systems might include carbon dioxide systems, clean agent systems, deluge sprinkler systems, and pre-action sprinkler systems. Underwriters Laboratories lists fire detection systems that can be used as releasing devices. Factory Mutual approves similar devices and systems. Fire detection systems can also be designed to close fire doors, shutters, and dampers, as well as open circuits to equipment, pressurize stairwell enclosures, and trigger venting mechanisms.

Fire detection systems can also automatically notify the fire department or an organization's fire brigade. This type of configuration can save valuable time responding to a fire. Configurations can range from a proprietary alarm that alerts occupants on the property to summon the fire department to a signal being transmitted to a centrally monitored station located off site. Off-site monitoring of the fire detection system allows the property to be constantly monitored, even when the property is isolated or unstaffed. Centrally monitored fire detection systems can help an organization realize cost reductions in insurance premiums (Factory Mutual Engineering Corp., 1971).

Several considerations should be analyzed prior to incorporating a fire detection system into a fire safety management program. First, the fire load, hazards, and probable fire characteristics should be evaluated. For example, is a fast-spreading, flammable liquid fire expected or a slow smoldering fire in stored rolls of paper? Next, activities normally conducted in the building should be analyzed to determine which ones might generate smoke or products of combustion. Considerations might include welding activities, waste incineration, or processes that create radiant surfaces. Third, the air flow within the building should be analyzed. Such an analysis can be helpful for properly selecting and placing detectors, especially smoke detectors. Fourth, a delay time for detection should be determined. The time necessary for evacuating the building, notifying fire fighters, or activating an extinguishing system should be computed. Finally, the cost of fire detection systems must be analyzed in order to provide a reliable, yet cost-effective, fire detection system (Grabowski, 1972).

RADIATION DETECTORS

Courtesy, Cerberus Pyrotronics

Flame detectors activate in response to radiant energy generated by the flame or combustion. Two types of flame detectors are commonly used. They include ultraviolet (UV) detectors and the infrared (IR) detectors. Figures 6–1 and 6–2 illustrate these detectors.

Figure 6–1. Ultraviolet Flame Detector

Infrared detectors operate at wavelengths above 7000 Angstrom. Therefore, they operate efficiently when separated from the flame by height and distance. Infrared flame detectors are often used in large, open, hazardous areas where there is a possibility for an immediate flame producing fire such as a flammable liquid fire. The wavelengths visible to the human eye range are between 4000 and 7000 Angstrom or from violet to red on the color spectrum. Flame detectors discriminate between fire-induced optical radiation and natural or artificial illumination. The ability to distinguish between fire and nonfire radiation is an essential feature of an effective flame detector. An infrared detector has a filter and lens system that screens out certain wavelengths and focuses infrared waves on a photoelectric cell. Infrared detectors respond within 50 to 100 milliseconds to a 1-square-foot fire at a distance of 25 feet.

Figure 6–2. Infrared Flame Detector

Courtesy, Cerberus Pyrotronics

The sensing element in an infrared flame detector varies but is usually a lead or cadmium sulfide cell or silicon solar cell. Infrared detectors are designed with time-delay units to give the unit time to determine the flicker frequency of the infrared radiation. Fires characteristically have flicker frequencies between 5 and 30 hertz. The time delay needed for determining the flame flicker frequency is between 1 and 30 seconds. This time delay allows for maximum ignition source and detector separation (Bryan, 1974).

The sensitivity of an infrared flame detector varies with design. Activation of the flame detector depends on the time delay, the response sensitivity, and the illumination intensity of the flames. The infrared flame detector alarm is activated when the level of illumination intensity of the flames reaches approximately 12 cycles per second.

The illumination intensity of diffusion flame combustion in a free-burning state is usually between 5 and 25 cycles per second. A general rule of thumb when evaluating infrared detector needs is that, as the wavelength spectrum of the sensitivity range is increased, the degree of response sensitivity is decreased. Artificial light sources such as incandescent lighting fixtures are usually modulated at about 120 cycles per second and therefore will not activate the detector (Bryan, 1974).

While broad-spectrum infrared detectors can respond to solar radiation, hot engines, reflections, lamps, and changes in humidity, they have also been effective in shielded locations such as vaults. The reliability of infrared detectors has been improved by filters within the detector that allow the device to focus on the radiation given off by hot carbon dioxide, a product of combustion. Some infrared detectors have a second sensor set for an intensity which helps differentiate between genuine fires and interfering sources. By comparing the two signals, reliability is increased and false alarms can be reduced (Larson 1985).

ULTRAVIOLET FLAME DETECTOR

The ultraviolet flame detector is designed to operate on the ultraviolet wavelengths below 4000 Angstrom. These wavelengths are primarily emitted by higher intensity flames. Most fires will have flames of sufficient intensity to produce wavelengths in the 2800- to 3000-Angstrom range. A major problem of ultraviolet flame detectors is that radiation from sunlight may be as low as 2900 Angstrom, which could cause false alarms (Bryan, 1974).

There are ultraviolet detectors that have been designed to respond to fire-induced optical radiation below 2900 Angstrom and not to respond to solar radiation at the same levels. One of these designs is the gas-discharge-type unit. This unit has been tested and found effective to the 2850-Angstrom level. Another effective type is the photosensitive silicon carbide diode. The photosensitive silicon carbide diode detector is very effective in enriched oxygen atmosphere at high pressure, but the detector still has difficulties with nonfire radiation. These types of detectors are excellent as explosion suppression system activators. One other type of ultraviolet detector that has been tested and found effective is the molybdenum crystal detector.

Ultraviolet flame detectors are designed and developed for specialized applications where the detector is relatively close to the expected ignition source. Ultraviolet flame detectors have been successfully used in explosion suppression systems and as releasing devices in hyperbaric chambers for water suppression systems. Because of the ultraviolet flame detector's sensitivity, the detector can be accurately set to respond to the ultraviolet wavelengths produced by the specialized expected ignition sources.

Ultraviolet and infrared flame detectors are sensitive devices and their placement should be based on an engineering survey of the conditions to be anticipated and the principle of operation. It is recommended that ultraviolet flame detectors not be placed near locations where arc welding, arcing from electric tool motors, germicidal lamps, or other sources of ultraviolet radiation are likely to be found. Also, infrared flame detectors should not be located near infrared lamps, matches, cigarette lighters, or other sources of infrared radiation, or where the ambient temperature is above 170° F.

THERMAL DETECTORS

The design of thermal detectors permits them to operate from thermal output or the heat from a fire. Laminar and turbulent convection air currents disperse the heat from the fire throughout the area. Turbulent air flow is induced and regulated by the fire plume thermal-column effect of rising heated air and gases above the fire surface. Knowledge of fire-induced thermal dispersion is important, because the rate of heat production from a fire and its distribution within an area are essential to the correct placement and operation of a thermal detector.

Fires produce heat products which are convected energy output, and visible and invisible particulate matter. The heat products warm the air surrounding a fire and the air expands, becoming more buoyant, and it begins to rise. The rising air forms a thermal column or fire plume. The fire plume rises to the ceiling where it is converted into a ceiling jet and radiates gas flow in various directions and distances. Air movement surrounding a fire is both vertical and horizontal. The fire plume characteristics and ceiling jet flow of turbulent convective heated gases are determined by the rate of heat release of the diffusion flame combustion and the height of the ceiling. There are many other factors in determining the placement of thermal detectors, but basic understanding of air flow is very important (Bryan, 1974).

There are two basic designs of thermal detectors and two other variations of these. The two basic designs are fixed-temperature and rate-of-rise detectors (Planer, 1979).

Fixed-temperature thermal detectors are highly reliable, stable, and easy to maintain, but are not very sensitive. There are two basic types of fixed-temperature thermal detectors. The spot detector (Figure 6–3) is a fixed-temperature thermal detector that involves a relatively small unit with a heat responsive element contained within the unit. When the fixed-temperature spot detector goes off, it must be renewed. Spot detectors are meant for *spot* detection or use in a small area.

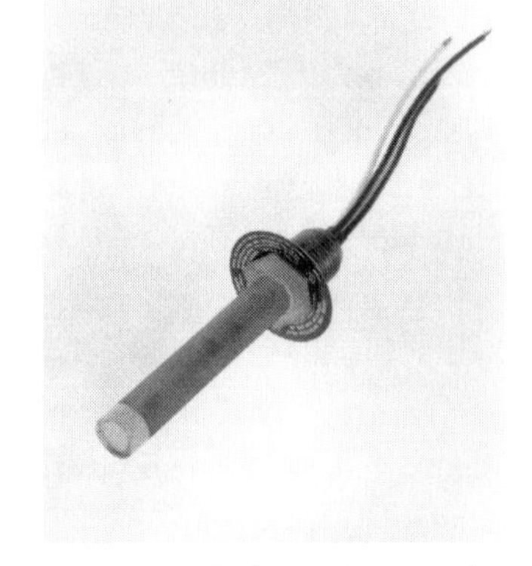

Figure 6–3. Thermal Fire Detector

Courtesy, Cerberus Pyrotronics

Another fixed-temperature detector is a line detector. The line detector is a thermal-reactive element located along a line of thermal-sensitive wire or tubing. Line detectors are electrically operated and activated by temperature. The line detector can be tested and reused after a fire if it has not been physically damaged by the fire (Grabowski, 1972).

Where a relatively fast fire is expected, the principal type of thermal detector that should be installed is that of a rate-of-rise detector. This detector operates when the fire plume raises the air temperature in the area at a rate above a specified threshold, usually at the rate of 15° F per minute. If the temperature of the fire does not exceed the threshold of the detector or develops slowly, the detector will not sense the fire.

Rate-of-rise units should not be placed where they will be affected by the building's heating system. Therefore, they are not recommended for use in warehouses, shipping areas, and hangars. Rate-of-rise detectors are often used in conjunction with fixed-temperature detectors for spot detection.

Rate-of-rise detectors are usually designed to operate either electronically or pneumatically. The rate-of-rise detectors that are operated pneumatically are often utilized as a releasing device for the operation of automatic extinguishing systems.

A new type of thermal detector is the rate-compensated detector. The rate-compensated detector is sensitive to both the rate of temperature rise as well as the fixed temperature. The rate-compensation detector is designed to eliminate the thermal lag found with a fixed-temperature detector, as well as the problem of false alarms and risk of missing slow heat-released combustions that plague the rate-of-rise detector.

SMOKE DETECTORS

Smoke detectors respond to the visible and invisible products of combustion. The visible products consist mainly of unconsumed carbon and carbon-rich particles. The invisible products consist mainly of solid particles smaller than five microns. There are two basic types of smoke detectors: the combustion products type and the photoelectric type.

The photoelectric smoke detector works by passing air through an enclosed unit with a light mounted at one end and the photoelectric cell at the other. As smoke particles are drawn into the unit and the reduced level of light intensity causes an unbalanced condition in the electrical circuit to the photoelectric cell, the detector is activated.

The projected or linear beam, photoelectric detector is one of the oldest, most established smoke detectors (Figure 6–4), and the linear smoke detector is capable of monitoring over long distances. The light beam may be projected over an area as much as 300 feet long. The detector utilizes infrared filters and modulated light to minimize extraneous light interference with the receiving unit. The detector operates when smoke blocks the light beam and reduces the light intensity received on the photoelectric cell. Projected beam detectors are very effective at detecting fires during their early stages (Bryan, 1974).

The reflected light beam detector and the projected beam detector are similar from the standpoint of operation. The reflected light beam detector can function as a *spot* smoke detector because of its very short light beam length. The design of the reflected beam

permits it to operate with a beam of light only two or three inches in length. A photoelectric cell mounted at right angles to the light sources, and a light catcher placed opposite the light source, constitute the source of lighting for the reflected beam type. The detector is activated by an increase in the intensity of light. This occurs when the light rays are reflected by the smoke into the photoelectric cell (Della-Giustina, 1979).

Figure 6–4. Linear Beam Smoke Detector

Courtesy, Cerberus Pyrotronics

Photoelectric detectors are relatively sensitive to smoke from smoldering fires but react rather slowly to flaming fires. Another drawback to photoelectric detectors is that they need an electrical source to operate, and therefore they are limited to where they can be placed.

The second basic type of smoke detector is the ionization-type, combustion products detector (Figure 6–5). The ionization-type detector detects both visible and invisible particle matter created by combustion. As mentioned earlier in the chapter, only particles five microns in size or larger are visible to the human eye. The ionization detector is most effective on particles from 1.0 to .01 microns in size. Most products produced by a fire are .01 to 1 micron in size.

The ionization detector works off either high voltage (AC), or low voltage (DC), depending on the design. The unit consists mainly of two electrodes and a sampling chamber in the area between the two electrodes. The oxygen and nitrogen air molecules in the chamber are ionized by alpha particles from the power source. The ionized pair of opposite-sign electrodes creates a minute, electrical current flow through the sampling chamber. When the particles from a fire enter into the chamber they reduce the mobility of the oxygen and nitrogen ions. This reduced ion mobility causes a reduction in the current flow and the detector is activated.

The ionization-type detector reacts to the velocity of ion mobility moving through the sampling chamber. Strong air currents other than from fire sources can cause the detector to activate when the sensitivity of units is high.

Figure 6–5. Ionization Smoke Detector

Courtesy, Cerberus Pyrotronics

Ionization-type detectors are designed for both spot detection and air duct detection. The air duct detectors are mounted either outside the duct area or inside the duct itself with an air shield attached to prevent false alarms. Because ionization-type detectors can operate off of low voltage, such as batteries, and can be placed almost anywhere, they are very popular for residential protection. One drawback is that ionization-type detectors respond well to a flame, but they are relatively insensitive to smoldering fires. Another type of smoke detector is a combination of two different types of smoke detectors, ionization type and resistance bridge type. The ionization, resistance bridge detector responds both to abnormal concentrations of primarily invisible particles of combustion. These are ionized by the radioactive source in the sampling chamber. The resistance bridge detector responds to the added water vapor in the air that is produced by a fire. This additional water vapor increases the conductivity of the electrical circuit on the glass plate and activates the detector. By combining the ionization detector and resistance bridge detector, a cumulative effect of both principles is achieved. A second chamber can be added to the ionization type and a second plate added to the resistance bridge type to compensate for normal ambient changes.

An engineering survey should be conducted to assess the conditions to be anticipated, and the principle of operation determined, before installing smoke-type detectors. The survey should provide for a graphic layout of the detector placement by blueprint design, and should comply with the detector's manufacturer-recommended location standards.

In conclusion, there are several factors a safety manager should look at when selecting a fire detector for a particular structure or location. As discussed earlier in the chapter, the following items should be considered:

1. Type of combustion to be expected (flaming or smoldering fires)
2. Activities normally conducted in the building that could generate smoke, heat, or flame
3. Air flow patterns within the area to be protected
4. Tolerable, detection delay time
5. Cost

Four other factors should also be considered in choosing a particular detector, and they are:

1. Reliability - the ability of the unit to function properly at all times
2. Maintainability - what and how much maintenance is required to assure optimal performance from the detector
3. Stability - the ability of the unit to sense fires over an extended period of time without a change in its sensitivity
4. Sensitivity - the time delay required by the unit in sensing a fire, and its activation, without false alarms

AIR SAMPLING DETECTOR

By the mid-1970s, it became necessary to compensate for the disadvantages of conventional fire detection technologies. Engineers in Australia created air sampling detection systems suitable for use in computer rooms, telecommunications facilities, offices, and residences. An air sampling system has an air transport system, filters to remove large dust particles, an optical detector to test the air, an air pump to move samples through the system, and a controller to interpret the detector results.

An air sample is drawn into the detector from the air transport system. Dust is filtered out to prevent contamination. The air sample is then exposed to a xenon light and drawn out by an aspirator. The light signal is transmitted to a photoelectric cell and passed on to a control card for processing onto a bar-graph representation of the smoke level. A stream of air is continuously drawn into and expelled from the detector (Lavelle, 1992).

Xenon lamps are extremely sensitive to a broad spectral band. The detector is able to respond to particles of all sizes. Xenon-based air sampling detectors are much more effective than conventional technologies in detecting fires in modern buildings, which are largely constructed from synthetic materials. Studies have shown that xenon-based air sampling detectors respond to a mass of smoke in the air, independent of particle size, and can be considered the most reliable and valid measures of fire intensity.

REFERENCES

Bare, W. K. *Fundamentals of Fire Prevention*. New York: John Wiley & Sons, 1977.

Betz, Gordon. "Legal Side of OSHA—What Does Safe Egress Really Mean?" *Plant Engineering*, October 12, 1978.

Brauer, Roger L. *Directory of Safety Related Computer Resources*. 2d ed. Des Plaines, IL: American Society of Safety Engineers, 1993.

Bryan, J. L. *Fire Suppression and Detection Systems*. Beverly Hills, CA.: Glencoe Press, 1974.

_____. "Human Behavior: A Critical Variable in Fire Detection Systems." *Fire Detection for Life Safety*. Washington, D.C.: National Academy of Sciences, 1977.

Bugbee, P. *Principles of Fire Protection*. Boston MA: National Fire Protection Association, 1978.

Building Officials and Code Administrators, Inc. *BOCA National Building Codes*. Country Club Hills, IL: Building Officials and Code Administrators, Inc., 1995.

Bush, L. S. *Introduction to Fire Services*. Beverly Hills, CA: Glencoe Press, 1970.

Canter, D. *Fires and Human Behavior*. New York: John Wiley & Sons, 1980.

Cellentani, Eugene N. "Let's Combine Detection, Alarm and Suppression." *Fire Journal* 82 (5) (1988): 13.

Clet, Vince H. *Fire-Related Codes, Laws and Ordinances*. Beverly Hills, CA: Glencoe Press, 1978.

Colburn, R. E. *Fire Protection and Suppression*. New York: McGraw-Hill Book Company, 1975.

Davis, Larry. "OSHA Standard for Fire Service." *Fire Engineering*, March 1981, pp. 27ff.

Della-Giustina, D. *School Safety World Newsletter*. National Safety Council, Chicago, Fall 1979.

Factory Mutual Engineering Corporation. *Factory Mutual Approval Guide 1972*. Norwood, MA: Factory Mutual Engineering Corporation, 1971.

______. *Handbook of Industrial Loss Prevention*. New York: McGraw-Hill Book Company, 1967.

Firenze, Robert. *The Process of Hazard Control*. Dubuque, IA: Kendall/Hunt Publishing Company, 1978

Grabowski, G. *Fire Detection and Activation Devices for Halon Extinguishing Systems*. Washington, D.C.: International Academy of Sciences, 1972.

Haessler, Walter M. *The Extinguishment of Fire*. Boston: National Fire Protection Association, 1974

Heskestad, G., and Yao, C. *A New Approach to Development of Installation Standard for Fire Detectors*. Norwood, MA: Factory Mutual Engineering Corporation, Research Corp., 1971.

Johnson, Joseph E. "Fire Detection Past, Present and Future." *Fire Journal* 81 (5) (1987):49–53.

Langdon-Thomas, G. *Fire Safety in Buildings; Principles and Practices*. New York: St. Martin Press, 1979.

Larson, T. E. "Detecting Fires with Ultraviolet and Infrared." *Specifying Engineer* 53 (5) (1985):62–65.

Lavelle, Laurie G. "Detecting Fires Before They Start." *NFPA Journal* 86 (3) (1992): 81–86.

Los Angeles Fire Department. *Fire Detection Systems in Dwellings*. Boston: National Fire Protection Association, 1963.

McGuire, J. H. *The Functions of a Fire Detection System*. Ottawa, Canada: National Research Council of Canada, 1967.

Merrit, Frederick S. *Building Engineering and Systems Design*. C. Litton Educational Publishing, 1979.

Mowrer, Frederick W. "Lag Times Associated with Fire Detection and Suppression." *Fire Technology* 26 (3) (1990):244–265.

National Fire Protection Association. *Fire Protection Handbook*. 17th ed. Quincy, MA: National Fire Protection Association, 1995

______. *Life Safety Code Handbook*. Quincy, MA: National Fire Protection Association, 1995.

______. *National Fire Codes*. Quincy, MA: National Fire Protection Association, 1995.

______. *NFPA 72: National Fire Alarm Code*. Quincy, MA: National Fire Protection Association, 1993.

Planer, R. *Fire Loss Control Management Guide*. New York: Marcel Dekker, 1979.

Robertson, J. C. *Introduction to Fire Prevention*. Beverly Hills, CA: Glencoe Press, 1975.

Underwriters Laboratories, Inc. *Fire Protection Equipment List*. Chicago, IL: Underwriters Laboratories, Inc., 1973.

Wells, B. *Fire and Theft Security System*. Blue Ridge Summit, PA: Lab Books, 1976.

STUDY GUIDE QUESTIONS

1. Name the four elements of a fire.
2. Describe the main attributes of an automatic fire detection system.
3. What are the limiting factors relating to the function(s) that an automatic fire detector can perform?
4. What are the components of an alarm/detection system?
5. List the three classes of automatic fire detectors and give examples of each class.
6. Differentiate between the three classes of fire detectors that will support their design purposes.
7. What is the slowest reacting type of fire detector? What is the fastest?
8. What factors should be identified prior to choosing an appropriate fire detection system?
9. What are the two types of flame detectors currently in use?
10. Briefly explain how an air sampling detector functions.

CASE STUDY

Fire Incident Event

Date of Accident:	December 31, 1994
Time of Accident:	9:00 PM
Location of Accident:	Petersburg Hospital, Petersburg, VA
Investigators:	Edward R. Comeau, Michael S. Isner
Losses Incurred:	Loss of five lives, major smoke damage

Summary Description of Event

The fire began in Room 418 apparently as a result of improper use of smoking materials, which ignited bedding that included an "air flotation" mattress with foam plastic padding. The fire intensified due to the damage to a wall-mounted oxygen regulator. When tested, it released oxygen at normal pressure which contributed to the fire's fast growth leading to the untenable conditions. The oxygen continued to flow until a maintenance person shut off a zone valve that allowed oxygen to flow to several rooms, including the patient's room. A nurse discovered the fire when the fire was already established in the patient's bed. The fire had not been detected sooner because there were no smoke detectors in any of the patient rooms. Smoke detectors were located in the corridor at 30-foot intervals. She had to leave the room to find a blanket to try to smother the fire. Upon returning to the room, she tried unsuccessfully to remove the patient from the room and put out the fire. The fire was growing too fast at this point, and she had to leave the room before being able to extinguish the fire or shut the door.

Smoke then spread out into the corridor because the door to the room where the fire began had not been closed. Smoke also spread into the noncombustible space above the ceilings of patients' rooms on the same side of the hallway as the room of the fire origin. The smoke entered these spaces because the rooms were not continuous from the floor to the underside of the floor above. Smoke spread from these spaces to other rooms on the same floor.

The switchboard operator initiated the emergency procedures, including a building-wide, coded announcement to the fire department. However, the fire department was not contacted until after the fire had broken the windows in the patient's room and was venting smoke outside. This rapid fire growth was a significant factor in the loss of life and property because it rendered the staff unable to successfully complete the emergency procedures.

The Petersburg Fire Department responded to the 911 call that was initially received by the Petersburg Police Department Emergency Communication Center, which is the Public Safety Answering Point (PSAP) for Petersburg. The call to the PSAP was answered within 30 seconds with the fire department dispatching two engines, a truck, ambulance, and a battalion chief. The battalion requested a second alarm response and the last two engines were dispatched. A neighboring community also sent a truck to the fire.

Damages caused by the fire included extensive smoke damage to the south wing of the hospital and the loss of five patients' lives due to smoke inhalation. The contents of Room 418 were completely destroyed.

Post-Response Assessment

Edward R. Comeau, Chief Fire Investigator, and Michael S. Isner, Senior Fire Investigator, investigated and analyzed the fire and concluded that the following contributed to the death of innocent lives and the destruction of property:

- Fire discovery was delayed.
- The fire alarm transmission to the fire department was delayed because the connection was taken out of service.
- The fire was already severe when it was discovered.
- Untenable conditions arose due to the rapid fire growth and the rapid development.
- The door between the room of fire origin and the corridor was left open.
- Walls between rooms were not continuous from slab to slab.
- There were no sprinkler systems in the room of origin or in the corridor.

In summary, the hospital staff acted in a manner consistent with actions that the *NFPA 101*, *Life Safety Code*, and *NFPA 1*, *Fire Protection Code*, required as part of a hospital's fire safety plan. These actions helped reduce the number of patients injured and/or killed as a result of the fire. The Petersburg Fire Department stated that, although everything had gone smoothly, they felt that making changes in fire protection hardware could have assisted in the extinguishing of the fire.

REFERENCE

Comeau, E. R., and Isner, M. S. *Hospital Fire Investigation Report*. Quincy, MA: National Fire Protection Association, 1994.

Fire Control Systems

CHAPTER 7

CHAPTER CONTENTS

FIGURES

The United States experiences more casualties and property loss due to fires than any other industrialized country. According to the National Fire Protection Association, 4050 were killed in fires in 1997. In this chapter, various types of fire control systems are identified and described. Automatic sprinkler systems, fire fighting foams, and chemical extinguishing agents are explained in order to assist safety managers with selecting an appropriate fire control system.

AUTOMATIC SPRINKLER SYSTEMS

In order to extinguish a fire, one of four methods can be used. These methods are:

1. Remove the heat.
2. Eliminate oxygen or dilute the oxygen concentration in the burning zone.
3. Remove the fuel.
4. Interrupt the chemical chain reactions.

There are different types of extinguishing agents and each can be effective when applied to fires for which they are best suited. When discussing fire control systems, the automatic sprinkler system should be discussed first. The automatic sprinkler system is considered basic protection. It is very versatile and can be installed to protect almost any hazard. Performance records maintained by insurers demonstrate that sprinklers are the most effective means of confining a fire to its place of origin. Many insurance companies recommend that automatic sprinklers be installed in buildings exceeding 5000 square feet. The U.S. General Services Administration installs sprinklers in all buildings exceeding five stories in height.

A sprinkler system, as well as any other type of fire control system, must incorporate three basic features for it to be considered a viable means of fire protection. It must (1) detect the presence of a fire and transmit an alarm to the building occupants, (2) confine the fire to the area of origin, and (3) activate without human intervention. The most important fire safety objective is to protect human life. A fire control system achieves this objective by detecting the presence of a fire and transmitting an alarm that warns occupants of a fire's presence. By warning the occupants, they have sufficient time to safely evacuate a building and summon the fire department. The second most important fire safety objective is to protect property. A fire control system achieves this objective by applying an extinguishing agent that confines the fire to the area of origin. This prevents additional property from being damaged by a fire. Last, fire control systems must be capable of activating during a fire without human intervention. Fire control systems that function without human intervention are termed automatic. Automatic systems, if properly installed and maintained, are highly reliable when required to activate.

Automatic sprinkler systems are the most common type of automatic fire control system. Sprinkler systems are complex systems composed of numerous components and assemblies. Despite their complexity, sprinkler systems consist of some basic parts: a water supply, a control valve, a piping system that distributes the water, and sprinkler heads that disperse the water. Sprinkler heads are threaded into the piping. They have an orifice that is plugged by a fusible link or quartzoid bulb. The link or bulb melts at predetermined temperatures. The link or bulb is thrown from the orifice by the water pressure. Water then free flows from the sprinkler orifice and is dispensed into a fan pattern by a deflector on the sprinkler head. The fanned water pattern removes heat from the fire and discharges water over the surface of surrounding combustibles. A common fallacy is that sprinkler systems discharge water from all of the sprinkler heads on the system, thus causing extensive water damage. Sprinkler systems actually discharge water from sprinkler heads that have activated over the fire. Therefore, water damage is minimized.

Sprinkler Heads

The standard sprinkler head provides, at a distance of 4 feet below the deflector, a discharge covering a diameter of approximately 16 feet when discharging at 15 gallons per minute. Orifice sizes vary but the minimum allowable orifice is 1/2 inch in diameter. Sprinkler heads can be installed in an upright or downward (pendant) position. Sprinklers are available with different temperature ratings. Sprinkler heads are color coded by their temperature rating in accordance with *NFPA 13* as shown in Figure 7–1.

Temp. Rating (° F)	Temp. Classification	Color Code	Glass Bulb Colors
135–170	Ordinary	Uncolored	Orange or Red
175–225	Intermediate	White	Yellow or Green
250–300	High	Blue	Blue
325–375	Extra High	Red	Purple
400–475	Very Extra High	Green	Black
500–575	Ultra High	Orange	Black
650	Ultra High	Orange	Black

Figure 7–1. *NFPA 13,* Temperature Ratings and Color Code

Maintaining adequate clearance below a sprinkler head is important to the effective operation of the sprinkler head. A minimum, 18-inch clearance should be maintained between the top of any storage and a sprinkler deflector. If a solid pile of stored material is 15 feet high, then there must be a 36-inch clearance maintained. If loose or palletized storage is 12 feet high, then a 36-inch clearance is required.

Types of Automatic Sprinkler Systems

The four basic types of automatic sprinkler systems are described below.

1. **Wet Pipe System.** The wet pipe system is utilized to protect property that is maintained at a constant temperature of at least 40° F. Wet pipe systems derive their

name because water is constantly maintained throughout the system. Therefore, wet pipe systems can only be utilized where temperatures will not reach the freezing levels. Wet pipe systems have the quickest reaction time, that is they can apply water to fire faster than any other system because water is maintained in the pipes.

2. **Dry Pipe System.** The dry pipe system is utilized to protect property that is susceptible to freezing temperatures. Water is not maintained in the distribution pipes. Instead air or nitrogen is maintained in the pipes under pressure. Thus, when a sprinkler head activates, the pressurized air or nitrogen in the system must be expelled before the water is applied. Dry pipe systems have a slower reaction time than wet pipe systems. Dry pipe systems must apply water from the activated sprinkler head within sixty seconds.
3. **Deluge System.** The deluge system is unique compared to the other types of sprinkler systems. The sprinkler orifices are kept open. The orifices are not plugged by a fusible link or quartzoid bulb. Deluge systems are used to protect high-hazard areas where a fast-spreading fire is anticipated due to the fire load. The deluge system is activated by a fire detection system connected to the sprinkler valve. When a fire detector senses a fire, it transmits a signal that opens the sprinkler valve. Water then flows from all of the sprinkler heads.
4. **Pre-action System.** The pre-action system is also unique. The pre-action system employs automatic sprinklers attached to a piping system containing air that may or may not be under pressure. A fire detection system is connected to the main sprinkler system valve. Actuation of the fire detection system causes the valve to open and permits water to enter the distribution pipes and be discharged through any of the sprinkler heads. The sprinkler heads activate at a temperature greater than the temperature setting of the fire detectors. The advantage of the pre-action system is that the delay affords occupants time to extinguish the fire with portable fire extinguishers or give another fire control system time to activate. Water is only applied if the human intervention or other fire control system fails to control the fire. This system is used to protect property that is sensitive to water damage, such as computer rooms.

NFPA 13, Standard for the Installation of Sprinkler Systems, provides the performance criteria and design specifications for all of the types of automatic sprinkler systems.

It is best to have an alarm that indicates water flow at the building and at another location, such as a central monitoring station. All valves controlling water supplies to sprinkler systems should indicate whether the valve is open or closed. Sprinkler system control valves should be closely monitored to prevent tampering. Approximately 30% of reported sprinkler failures are due to closed valves.

One of the most important requirements of automatic sprinkler systems is adequate maintenance. This includes periodic inspection and testing. *NFPA 25, Standard for the Inspection, Testing and Maintenance of Water-Based Fire Protection Systems*, outlines recommended maintenance frequencies and procedures.

Water Supply

Adequate water supplies are needed for automatic sprinkler systems. Sprinkler systems can be supplied by a variety of methods including a public main, a private storage tank and fire pump, or a private pond or lake and fire pump. Sprinkler systems also incorporate a fire department connection for a fire department pumper to supplement the water supply and pressure if needed. Whatever water supply method is chosen, water supplies must be evaluated prior to designing and installing sprinkler systems. A water supply should only be measured by a professional trained for water supply testing. The water supply evaluation should include a review of the public water supply, including maps of the distribution system showing locations of mains and valves; records of consumption, storage levels, gate valve and hydrant inspections; and an actual measurement of water supply output from the nearest fire hydrants. This information is gathered and a fire flow is calculated. Fire flow is the amount of water available for fire fighting purposes.

The water supply required for a sprinkler system is termed the *demand*. A system's demand consists of the water required for sprinkler systems at the most hydraulically remote area of the system, plus interior hose lines, and exterior hose lines. The available fire flow must be larger than the system demand.

CARBON DIOXIDE SYSTEMS

Carbon dioxide extinguishes primarily by displacing part of the atmosphere at or near the fire so that the oxygen supply in the vicinity is reduced to at least 15% by volume. Carbon dioxide has little cooling effect and is therefore not the best material to use on deep-seated fires. Another drawback is that carbon dioxide systems provide only limited quantities of extinguishing agent and, consequently, must extinguish immediately unless provided with a reserve supply. While we generally think of sprinkler systems in terms of basic protection, a carbon dioxide system is specifically engineered for a special hazard. Carbon dioxide is limited to extinguishing Class A-, B-, and C-type fires. *NFPA 12, Standard for Carbon Dioxide Extinguishing Systems*, contains the performance criteria and design specifications for carbon dioxide systems.

Carbon dioxide's use as an extinguishing agent began during World War I. Further development and improvements followed rapidly. The main advantage of its use is that it can be easily converted to a gas from pressurized liquid storage. Discharge is accomplished by conducting the liquefied gas through a dip tube. Dense clouds of carbon dioxide are formed when it is discharged to the atmosphere. Its application ranges from small portable fire extinguishers to large multi-ton systems (Colburn, 1975).

Carbon dioxide is primarily used for flammable and combustible liquid fires and electrical equipment fires. Carbon dioxide is mildly toxic and can cause a person to become unconsciousness if trapped in a 9% concentration. Since carbon dioxide is a nonconductor, it is effective for controlling fires in electrically energized equipment. However, along with the other limitations mentioned, hot embers may rekindle after the carbon dioxide gas has been dissipated and no longer affords a smothering effect. Carbon dioxide has been used successfully on flammable liquid fires, and in particular, on air-crash fires when a quick knockdown is needed (Haessler, 1974).

Certain metals can decompose when exposed to carbon dioxide. Another limitation is that, when compared to sprinkler systems, the reliability is considerably less because of frequently experienced component malfunctions.

Carbon dioxide offers many advantages. It is colorless and odorless. Even though it is stored in a liquefied form, it gasifies leaving no residue. Therefore, it can be considered a clean agent. These systems can be categorized into high-pressure or low-pressure systems. They can also be applied either by total flooding or local application. Low-pressure systems are generally considered more economical if the volume to be flooded exceeds 2000 cu. ft. (Haessler, 1974). A total flooding system consists of a supply of carbon dioxide connected to fixed piping with nozzles arranged to discharge carbon dioxide into an enclosure around the hazard, as opposed to a local application system.

Carbon dioxide extinguishing requirements will depend on the location of hazards to be protected and the position of the discharge nozzles. In designing a system, the engineer would want to refer to UL's *Fire Protection Equipment Directory*. This is a listing which provides information regarding the specified configurations of various types of nozzles in relation to hazards, flow rates in pounds per minute, area of coverage, and minimum and maximum quantities per nozzle.

Another important component of the carbon dioxide system is the type of detection device to be utilized. Detection devices should be interlocked to alarms in such a way as to immediately react when heat, smoke, flame, or combustible vapors are identified. The type of detection device will depend upon the hazard (Bryan, 1974).

As previously stated, carbon dioxide has become very widespread in use because of its low cost, wide availability, cleanliness, effectiveness, and nonconductive properties. Because carbon dioxide is discharged in a gaseous form by internal storage pressure and the vapors are heavier than air, carbon dioxide extinguishing systems are usually found in interior locations, protecting the following hazards (Whitman, 1979):

- Flammable and combustible liquids
- Electrical hazards including transformers, oil reservoirs, switches, circuit breakers, motors, and generators
- Sensitive electronic equipment including computers, printers, and radio or television transmitting equipment
- Food preparation areas such as deep fat fryers, ovens, broilers, ranges, and exhaust ducts
- Internal combustion and diesel engines and accompanying fuel supplies
- Combustible materials with unique intrinsic value such as legal documents, official records, furs and films, and other items often stored in vaults

Carbon dioxide systems are not to be installed for the protection of the following hazards:

- Occupied areas which may not be evacuated prior to carbon dioxide discharge
- Chemicals or pyrophoric materials containing an inherent oxidizing agent
- Reactive metals including sodium, potassium, magnesium, titanium, zirconium, and the metal hydrides

It is extremely important when operating a carbon dioxide system that personnel protection is considered. For example, if a total flooding system were in operation it would be necessary to provide predischarge alarms and delay devices that would allow sufficient time for personnel to evacuate the area before the gas is discharged.

FOAM EXTINGUISHING SYSTEMS

Foam extinguishing systems have been exclusively used for many years, especially in the petro-chemical industry, for the extinguishment of flammable and combustible liquids fires. Two kinds of foams are primarily used today. They are chemical and mechanical foams, depending upon how they are generated. *NFPA 11, Standard for Low Expansion Foam and Combined Agent Systems*; *NFPA 11A, Standard for Medium and High Expansion Foam Systems*; and *NFPA 11C, Standard for Mobile Foam Apparatus*, can be referenced for specific performance criteria and design specifications.

In the past, chemical foam systems for large oil storage tanks consisted of two tanks of stored solutions which, when pumped through a piping system to a mixing chamber, applied foam on top of the burning fuel. Chemical foam systems have functioned through the use of either single or double powder generators; the A powder (aluminum sulfate) in water solution with B powder (bicarbonate of soda) could form a continuous blanket on the surface of the flammable liquid and separate the combustible vapors from the oxygen necessary for combustion. Due to the fact that the bubbles are of lower specific gravity than oil or water, they float on the liquid surface, thus excluding oxygen from the surface causing extinguishment. In more recent years, mechanical foam, which is also called air foam, has come into widespread use. Mechanically generated foams were developed in Germany in the 1930s which provided for concentrates to be introduced into water and mechanically expand it with air bubbles. The foam concentrate is made of hydrolyzed animal or vegetable protein, stabilizers, solvents, and an industrial germicide.

These regular protein-type mechanical foams are suitable for ordinary hydrocarbon liquids and are available for proportioning in water in 3% or 6% concentrations. Protein-type foams are biodegradable, nontoxic, noncorrosive, and present no major cleanup problems. They also have a good resistance to burn-back, good extinguishment, and can be obtained at a relatively low cost. There are four methods of producing air foam: nozzle aspirating systems, in-line foam pump systems, in-line aspirating systems, and in-line compressed air systems. The names of these systems indicate where and how air is injected into the water-concentrate solution to produce air foam.

In today's broad spectrum of chemicals, there are certain flammable liquids that have the capability of deteriorating foams. The type of solvents which can destroy the regular protein-type foams are called polar or alcohol-type solvents. These solvents are miscible with water and other constituents of the foam. The solvents, in particular, are alcohols, ether, and ketones. A special type of foam called aqueous film-forming foam (AFFF) combined with dry chemicals can be used to extinguish these alcohol-type solvent fires.

The development of AFFF has been fairly recent and it has proven to be a superior foam in many cases. There are two different types of aqueous film-forming foam. One type of AFFF is a protein-based foam and the other is an alcohol-resistant foam. The protein-based foam concentrates are suitable for extinguishing hydrocarbon fuel fires only and provide excellent heat resistance and post-fire security. The foam blanket excludes oxygen from the fuel's surface, and the water in the foam provides cooling.

The alcohol-resistant AFFFs are the most flexible and widely used foams. They are used for extinguishment and vapor suppression of hydrocarbon and polar solvent (alcohol) fires, fires from a mixture of these fuels, and fires from oxygenated motor fuels. These foams, which are fluorinated surface-acting agents, have very fast knockdown capabilities and are extremely effective when used on fuel spills. They have been found to be 25% to 30% more effective than the protein-type foams.

AFFFs can also be used in conventional closed-head sprinkler systems. One disadvantage of the AFFFs is that they are fast draining, so they do not provide long-term effectiveness as the other foams do.

Foams may be applied to fires through fixed systems. Fixed foam extinguishing equipment may be automatic, manual, or a combination of both. Some systems consist of one or more portable foam extinguishers suspended in such a way that flame or heat releases a fusible link so that the extinguisher tips over for automatic operation. Other systems, which are more complex, consist of fixed piping through which the foam-producing solutions move to a number of deflectors onto the fire surface. Such a system can be operated either manually or automatically by the use of heat-sensitive actuators, and varies in the discharge rate from 15 to 4000 gallons per minute. These fixed systems are most often used to protect dip tanks, oil and paint storage rooms, asphalt coating tanks, and other tanks which are used for storing large capacities of flammable liquids. The diameter of the tank determines the number of foam discharge outlets.

The foams discussed so far have been those with low-expansion levels. Low expansion means having expansion ratios of 0–20:1. There are also high-expansion foams which provide a foam-to-solution ratio of 200:1 to 1000:1. These high-expansion foams are defined as an aggregation of bubbles from mechanical expansion of a foam solution by air or other gases. This type of foam can be used in total flooding systems designed to fill enclosed spaces such as rooms or buildings.

Foam is also effective in fighting fires in inaccessible places such as coal mines and building basements. High-expansion foams extinguish by excluding oxygen from the air when the water of the foam is converted to steam and by the cooling effect when the water converts to steam and absorbs the heat from the fuel. This type of foam also has the characteristic of insulating materials that are not involved in the fire. High-expansion foam is generated by forcing the foam concentrate through a mesh screen using large volumes of air. The foam produced is then applied directly on the fire through the use of ducts or chutes.

After the foam has been dispersed on the fire, it should be allowed to soak into the fuel for at least one hour. Following the soaking period, firefighters may then proceed to cut through the foam by using a water spray to investigate more thoroughly. The one disadvantage that foam has is that it does impair vision if it is higher than head level. Special care must also be taken when using it on electrical equipment. Always make sure that the equipment is de-energized before any application. Foams do have their place in today's industrial facilities. There are various types of foams available and different methods of generating them. If the foams are used correctly for their intended purpose, they can prove to be a valuable tool in controlling, extinguishing, and confining fires in the workplace.

HALON EXTINGUISHING SYSTEMS

At the beginning of World War II, it became necessary to create nonconductive fire extinguishing agents for use on aircraft and ship fires. Before 1940 carbon tetrachloride was widely used for these applications. Carbon tetrachloride was removed from service because it causes cancer. In 1939–40, Germany developed chlorobromomethane, commonly known as Halon 1011, to replace methyl bromide (Halon 1001). Methyl bromide was also highly toxic.

Halogenated extinguishing agents are hydrocarbons in which one or more of the hydrogen atoms have been replaced by atoms of the halogen family. The halogen family includes fluorine, chlorine, bromine, and iodine. Substituting a halogen for one or more hydrogen atoms makes the compound nonflammable and desirable as an extinguishing agent. Halogen extinguishing agents were widely used in portable fire extinguishers and fire suppression systems (Cote and Bugbee, 1988).

In 1947, the U.S. Army Chemical Center and Purdue University tested the extinguishing properties of 60 agents on electrical, petroleum, and engine compartment fires. The halons were evaluated for their nonconductivity, density (compactness of storage), and corrosive or abrasive residues. Halons extinguish fires by interrupting the chemical chain reactions. Halons containing bromine were found to be much more effective than those with chlorine or fluorine. Halon 1202, dibromodifluoromethane, was found to be the most effective and most toxic. Halon 1301, bromotrifluoromethane, was the second most effective and least toxic. Between 1960 and late 1980, halons were used in total-flooding fire suppression systems, particularly in libraries and computer rooms. These systems were popular because the halons left no residue in these sensitive areas. Extensive testing in the sixties and seventies demonstrated the value of Halons 1301 and 1211.

The 1947 Army/Purdue tests revealed that at temperatures above 900° F halons decompose. When halons decompose, they release hydrogen fluoride, hydrogen bromide, free bromine, and carbonyl halides. These byproducts were found to be harmful if inhaled.

Halons also remain in the atmosphere for long periods. The atmospheric life of Halons 1211, 1301, and 2402 is over 300 years. As these compounds reach the stratosphere (15–50 km above the earth), they release chlorine and bromine. These elements have been found to deplete the ozone layer in a similar manner as chlorofluorocarbons (CFCs). CFCs contribute to the breakdown of the earth's ozone layer. The earth's ozone layer filters out ultraviolet radiation from the sun. Scientists attribute a hole in the ozone layer to increased production of CFCs and similar products. The increased ultraviolet radiation reaching the earth's surface can lead to harmful health effects with chronic exposure. Prolonged exposure can result in melanomas, cataracts, and immune system failure, as well as altering aquatic and terrestrial ecosystems.

In 1987, the United States was one of 24 nations to sign the Montreal Protocol on Substances that Deplete the Ozone Layer. The document called for production and consumption of Halons 1211, 1301 and 2402 to be frozen at 1986 levels. Production and consumption were to be cut in half by 1998. Beginning in 1990, and at least every four years thereafter, on the basis of the latest scientific, technical, and economic information, new chemicals could be added or dropped. Also, additional uses could be banned and the phaseout schedules tightened. In the United States, the Environmental Protection Agency (EPA) has enacted additional rules regulating the production, use, handling, and depositing of halons. In 1989, the Omnibus Budget Reconciliation Act increased the federal excise tax on halons.

The U.S. government subsidized the Halon Alternatives Research Corporation. It examined the critical properties of halons and determined criteria for halon substitutes. Substitutes must satisfy the following criteria:

- fire suppression efficiency
- low residue levels
- nonconductivity
- low metals corrosion
- high materials compatibility
- stability under long-term storage
- low toxicity of the chemical and its decomposition products
- low stratospheric ozone depletion potential

Halon substitutes were first intended to be used as drop-in replacements in existing halon fire suppression systems. Initial research revealed that most substitutes had characteristics that did not allow them to serve as drop-in replacements. Halon substitutes often required higher extinguishing concentrations, lower volumes, higher vapor pressures, and lower, maximum discharge times.

The NFPA responded in 1994 by publishing a consensus standard for clean halon alternatives. *NFPA 2001, Standard on Clean-Agent Fire Extinguishing Systems*, details the requirements for design, installation, testing, and maintenance of fire suppression systems utilizing clean agents.

HALON ALTERNATIVES

Several alternatives are commercially available to replace halon extinguishing agents. The alternatives include new chemical extinguishing agents, inert gases, water mists, and powders.

Commercialization of clean-agent fire suppression systems was fostered by many organizations. The EPA identified acceptable clean agents in its Significant New Alternatives Policy (SNAP) program under the provisions of the Clean Air Act. The EPA regularly tests new chemicals for various industries to assure that they are environmentally friendly. The EPA's SNAP list of alternatives is highlighted in Figure 7–2.

Total Flooding Agents for Occupied Areas	
C_4F_{10} (PFC-410 or CEA-410)	IG-55 (Argonite)
C_3F_8 (PFC-218 or CEA-308)	IG-541 (Inergen)
HCFC Blend A (NAF S-III)	Water Mist
HFC-23 (FE13)	Carbon Dioxide
HFC-227ea (FM-200)	Sprinklers
IG-01 (Argon)	
Total Flooding Agents for Unoccupied Areas	
CF_3I	HCFC-22
HBFC-22BI (FM-100)	HCFC-124
HFC-125	Inert gas/powdered aerosol blend (FS 0140)
Gelled halocarbon/dry chemical suspension (PGA)	Powdered aerosol A (SFE)
Miscellaneous Total Flooding Agents	
Sulfur Hexafluoride (SF6)	
Note: allowed for only military use and civilian aircraft.	
Streaming Agents	
C_6F_{14}	Gelled halocarbon/chemical suspension
HCFC Blend B (Haoltron 1)	Water Mist
HCFC Blend C (NAF P-III)	Carbon Dioxide
HCFC Blend D (Blitz III)	Dry chemical
HCFC-123	Water
HCFC-124 (FE-241)	Foam

Source: Seaton, 1995.

Figure 7–2. Clean Agents Accepted by EPA SNAP Program

WATER SPRAY SYSTEMS

Water spray systems are designed to protect various types of hazards which might include flammable and combustible liquids. These systems may be designed to be automatic or manually operated. Water spray systems are not widely used for the general fire protection of buildings. They are used to protect chemical processing equipment, tank car/tank truck facilities, and storage tanks containing hazardous liquids or gases, transformers, and other units. They are particularly useful in situations where it could be difficult to utilize manual hose streams. The design of water spray systems is highly specialized. These systems extinguish fires by cooling, smothering, and dilution. Water spray systems must be able to operate for the time necessary to provide cooling, dispersing, and dilution. In addition, water sprays are compatible with foam, thus allowing both agents to be used simultaneously. *NFPA 15, Standard for Water Spray Fixed Systems for Fire Protection*, contains the requirements for the design, installation, testing, and maintenance of water spray systems.

DRY CHEMICAL EXTINGUISHING SYSTEMS

Dry chemicals are mixtures of specially formulated powders. They are effective on Class A, B, and C fires, especially on flammable liquids and grease fires. Dry chemicals are effective at inhibiting chemical chain reactions within the fire. They also provide a small smothering effect. Dry chemical agents are commonly used in portable fire extinguishers, wheeled fire extinguishers, hand hose line systems, and engineered systems. *NFPA 17, Standard for Dry Chemical Extinguishing Systems*, contains the requirements for design, installation, testing, and maintenance of dry chemical extinguishing systems.

Sodium bicarbonate is a commonly used dry chemical. It is an excellent extinguishing agent for flammable liquids and electrical fires. One disadvantage of sodium bicarbonate is its corrosiveness. It can affect finely polished metal surfaces usually found in electronic and computer systems. Potassium bicarbonate, another dry chemical, is also suitable for flammable liquids.

Monoammonium phosphate is considered the multipurpose, dry chemical agent. It is suitable for ABC-type portable fire extinguishers. It is very durable and it is especially desirable for use in fire extinguishers that will be used by untrained personnel. For this reason, it eliminates a decision on the part of the user to use a suitable agent on a particular fire. Dry chemicals also have limitations. They are not effective on materials that contain their own oxygen supply such as cellulose nitrate. Dry chemical extinguishers should not be used on fires involving combustible metals such as sodium, magnesium, titanium, potassium, and zirconium.

PORTABLE FIRE EXTINGUISHERS

Portable extinguishers are probably the most common of all private fire protection equipment. Portable fire extinguishers have a limited supply of fire extinguishing agent. These extinguishers should be maintained in a fully changed and operable condition, and kept in their designated places at all times when not being used. In addition, the extinguishers should not be obstructed and should be stored in clear view of those who might need to use them. Such extinguishers are usually thought of as portable because of their relatively small weight and bulk; however, larger portable extinguishers are provided with wheels which permit the apparatus to be moved by one or two persons. These extinguishers should be relied upon only to the extent of their intended use, and when that limit has been reached, large fire fighting equipment should be used. *NFPA 10, Standard for Portable Fire Extinguishers*, contains the requirements for various types of fire extinguishers.

Portable fire extinguishers are classified by nationally recognized testing laboratories for use on certain classes of fires and rated for relative extinguishing effectiveness at a temperature of +70° F. They are also effective on small fires in electrical equipment such as panel boards, switch-boards, motors, and other Class C fires where a nonconducting extinguishing agent is of importance. Dry chemical extinguishers are not suitable for seated fires in ordinary combustible materials such as wood, paper, textiles, and other Class A fires, which require the cooling effect of water for complete extinguishment. The extinguishers may be of some value for surface fires in small quantities of material where the smothering effect of the extinguishing agent may be effective.

A nonpressurized extinguisher is operated by pushing the handle down which punctures a sealed disc in the cartridge. The released gas pressurizes the dry chemical chamber and expels the dry chemical. The discharge is controlled by the shut-off nozzle at the end of the hose. In the pressurized dry chemical extinguisher, both the dry chemical and expellent are stored in a single chamber under a pressure of about 150 psi. The nozzle may be opened allowing the stored air pressure to expel the dry chemical from the chamber through the hose by squeezing or gripping the extinguisher nozzle handle. Release of the extinguisher nozzle handle provides a shut-off feature (Bare, 1977).

Ready-for-service fire extinguishers depend upon:

1. people trained in the use and handling of extinguishers
2. suitable location of extinguishers
3. sound working order of extinguishers
4. suitable types of extinguishers for hazards encountered
5. early warning of the fire for the fire extinguisher to be effective

REFERENCES

American National Standards Institute. *Responding to Hazardous Material Incidents.* New York: American National Standards Institute, 1993.

Bare, William K. *Fundamentals of Fire Prevention.* New York: John Wiley & Sons, 1977.

Bryan, J. L. *Automatic Sprinkler and Standpipe Systems.* Boston, MA: National Fire Protection Association, 1979.

______. *Fire Suppression and Detection Systems.* Beverly Hills, CA: Glencoe Press, 1974.

Building Officials and Code Administrators, Inc. *BOCA National Building Codes.* Country Club Hills, IL: Building Officials and Code Administrators, Inc., 1995.

Bush, Loren S., and McLaughlin, James. *Introduction to Fire Science.* Beverly Hills, CA: Glencoe Press, 1970.

Colburn, Robert E. *Fire Protection and Suppression.* New York: McGraw-Hill Book Company, 1975.

Cote, Arthur, and Bugbee, Percy. *Principles of Fire Protection.* Quincy, MA: National Fire Protection Association, 1988

Fire Protection Systems. *Fire FM-200 Clean-Agent Fire Extinguishing Systems.* Blue Springs, MO: Fire Protections Systems, 1993.

Glenn, William. "Canada Moves Quickly on CFCs and Halons." *Occupational Health and Safety Canada*, May/June 1993, pp. 22–26.

Haessler, Walter M. *The Extinguishment of Fire.* Boston: National Fire Protection Association, 1974.

Harrington, Jeff L. "The Halon Phaseout Speeds Up." *NFPA Journal*, March/April 1993, pp. 39–42.

Merritt, Frederick S. *Building Engineering and Systems Design.* C. Litton Educational Publishing, 1979.

National Fire Protection Association. *Fire Protection Handbook.* 17th ed. Quincy, MA: National Fire Protection Association, 1994.

______. *National Fire Codes.* Quincy, MA: National Fire Protection Association, 1995.

______. *National Fire Codes, 2001: Standard on Clean-Agent Fire Extinguishing Systems.* Quincy MA: National Fire Protection Association, 1994.

North American Fire Guardian Technology, Inc. *Product Summary of Fire Extinguishing Agent NAFS-III (HCFC BLEND "A").* Vancouver: North American Fire Guardian Technology, Inc., 1993.

Planer, Robert G. *Fire Loss Control: A Management Guide.* New York: Marcel Dekker, 1979.

Robertston, Loslie E., and Naka, Takeo. *Tall Building Criteria and Loading.* New York: American Society of Civil Engineers, 1980.

String, Clyde, and Irvan, Rick. *Emergency Response and Hazardous Chemicals Management.* Delray Beach, FL: 1993, n. p.

Underwriters Laboratories, Inc. *Fire Protection Equipment Directory.* Northbrook, IL: Underwriters Laboratories, Inc., 1981.

Whitman, Lawrence E. *Fire Prevention.* Chicago, Illinois: Nelson-Hall, 1979.

Woodside, Gayle. *Hazardous Material and Hazardous Waste Management.* New York: John Wiley & Sons, 1993.

Workplace Safety in Action: Hazard Assessment. Neenah, WI: J. J. Keller and Associates, 1993.

STUDY GUIDE QUESTIONS

1. What is the difference between a wet pipe system and a dry pipe system?
2. High-expansion foam extinguishes fire in three ways. What are they?
3. How do you determine when maintenance is necessary to keep fire protection equipment functional?
4. AFFF is a special type of foam. What is it and on what type of fires is it used?
5. What are the three areas that need to be monitored in pressure tanks? Explain their importance?
6. List the temperature ratings for automatic sprinklers and explain why one single rating cannot be used in all situations.
7. Why is halon being phased out as a fire suppressant?
8. List and discuss some alternatives to halon.
9. What is the most common form of private fire protection equipment?
10. List the different classes of fire extinguishers.

CASE STUDIES

1. A moving company is storing residential contents and commercial documents in a 75,000-square-foot warehouse facility. Justify the installation of automatic sprinkler protection by demonstrating the cost reduction on insurance premiums. Demonstrate the cost reduction for a 10-year period.
2. A safety manager receives budget approval for $320,000 over a 5-year period to install automatic sprinkler protection in 17 buildings. From a risk management standpoint, develop a detailed strategy for determining which buildings should receive sprinkler protection first and throughout the 5 years.
3. A sprinkler system must be removed from service for maintenance. The sprinkler system protects an industrial oven that is not operating due to a plant-wide shutdown. Hot work is planned for the industrial oven. Develop an action plan for conducting the hot work safely during the sprinkler system's impairment.

Care, Maintenance, and Inspection

CHAPTER 8

CHAPTER CONTENTS

FIGURES

One critically important component of a fire safety management program is maintaining fire protection systems. Proper maintenance requires that all fire protection systems be inspected, tested, and maintained at frequencies recommended by manufacturers and the National Fire Protection Association (NFPA). This chapter outlines some crucial areas of maintenance and inspection for automatic sprinkler systems.

CARE AND MAINTENANCE OF SPRINKLER SYSTEMS

Sprinkler system reliability is dependent on an organization properly caring for and maintaining the system. Systems are not effective just because they are installed. Sprinkler systems are subject to failure and impairments due to many factors, including the surrounding environment, human error, mechanical breakdown, and neglect. Preventive maintenance must be performed on a regular basis to assure that sprinkler systems will operate as designed. If sprinkler systems are to be effective, then the system as a whole must be maintained. Sprinkler systems are composed of hundreds of components and assemblies that are interdependent on one another. Sprinkler system maintenance should be based on systematic analysis of the components and assemblies from the water supply to the sprinkler heads. *NFPA 25, Standard for the Inspection, Testing, and Maintenance of Water-Based Fire Protection Systems*, outlines the maintenance frequencies and good practices for caring for sprinkler systems.

Diligent and thorough attention to the care and maintenance of sprinkler systems is necessary to assure that the installed fire protection is reliable. The primary purpose of an automatic sprinkler system is to protect life and property. Serious fires seldom occur in properties completely protected with properly maintained, automatic sprinkler systems. Not properly caring for and maintaining sprinkler systems can create a false sense of security to an organization. Providing automatic sprinkler protection is typically a sign of good business foresight. The quality of business administration and the intelligence of management are reflected in the rigor of provisions made for maintaining sprinkler systems.

MAINTENANCE AS A FACTOR IN SPRINKLER SYSTEM PERFORMANCE

Actual experience data has shown the excellent performance of automatic sprinkler protection in the control and extinguishment of fires. NFPA statistics demonstrate that 96.2% of sprinkler systems successfully perform as designed. Numerous fires have been extinguished by the activation of one or two sprinkler heads and limited these potentially catastrophic fires to slight losses. Because only a small fraction of experience data is reported to the NFPA, sprinkler system performance may be better than indicated by the NFPA.

The responsibility for the prevention, discovery, or elimination of maintenance deficiencies may extend to a number of individuals in any one organization. Some deficiencies can be corrected during routine inspections. Others might not be so readily observed and would be discovered during systems tests by qualified persons. Still other deficiencies are due to poor management. This demonstrates the need for a consistent preventive maintenance program and a willingness to assign responsibility and accountability. In addition, it is necessary that effective cooperation exist with fire departments, insurance carriers, and other groups having occasion to provide outside aid in case of emergency.

RESPONSIBILITY FOR MAINTENANCE

Owners or managers of properties are responsible for maintaining automatic sprinkler systems in reliable condition. Owners and managers are responsible for the lives of those who are housed or employed on a property, as well as the continuity of production and employment. In order to obtain favorable results, proper attention must be given by owners and managers to the reliability of sprinkler systems.

Some major sprinkler system problems have been due to a lack of responsibility rather than lack of knowledge. Regardless of fire prevention inspections that may be made by others such as insurance carriers, fire departments, and service contractors, management alone must act to assure that fire protection features are in good operating condition at all times. Management must also assure that employees are trained to handle any fire situation effectively. To provide the necessary assurance, a preventive maintenance program should be developed and implemented by management.

A preventive maintenance program should identify what systems, assemblies, and components will be inspected, tested, and maintained, as well as when the actions should be taken. A sample inspection report form is shown in Figure 8–1. Maintenance procedures will have some seasonal differences, most of which are indicated later in this chapter under the various items of sprinkler system components being considered. The following are examples of inspections governed by seasonal effects.

Fall Inspection

At the approach of freezing weather in the fall months, an inspection should place special attention to several items. Cold-weather valves and drain pipes exposed to freezing temperatures should be closed (drain valves on the exposed piping are left slightly open). The specific gravity of antifreeze solutions in sprinkler systems should be tested. Dry pipe valves should be checked to make sure that they are holding air properly and that the electric and water-flow alarms are in order (the drains at low points of the dry piping should be checked to make sure that they are properly clear of water).

Fire Protection Equipment Inspection Report

Facility: ______________________ Inspector: ______________________

Location: ______________________ Date: ______________________

The Following Items Should Be Checked At Least Weekly.

Any "No" response should be explained.

WATER SUPPLY, SECTIONAL, AND SPRINKLER SYSTEM CONTROL VALVES

Valve ID	Open	Shut	Sealed	Valve ID	Open	Shut	Sealed	Valve ID	Open	Shut	Sealed	Valve ID	Open	Shut	Sealed

PUBLIC WATER

Public water supply in service? ☐ Yes ☐ No ______________________ Pressure: ________ psi

Fire department connection accessible, caps in place, couplings free to rotate? ☐ Yes ☐ No ____________

FIRE PUMPS

Pump ID	Type	Set For Auto.?		Operated Today?		Checklist Completed?		Comments
		Yes	No	Yes	No	Yes	No	

WATER SUPPLY TANKS

Tank ID	Tank Full?		Heater Working?		Water Temp.	Comments
	Yes	No	Yes	No		

AUTOMOTIVE FIRE APPARATUS

Each fully in service? ☐ Yes ☐ No ______________________

Checklist completed? ☐ Yes ☐ No ______________________

SPECIAL EXTINGUISHING SYSTEMS

System ID	Type	In Service?		Date Last Serviced	Date Last Tested	Comments
		Yes	No			

The Following Items Should Be Checked At Least Monthly.

Any "No" response should be explained.

WET PIPE, DRY PIPE, DELUGE, AND PRE-ACTION SPRINKLER SYSTEMS

System ID	Alarm Tested?		Water Pressure			Heat Adequate?		Air/ Supv. Press.	Comments
	Yes	No	Static	Flow	Differ-ential	Yes	No		

N127 Ed 4/82

Courtesy, Industrial Risk Insurers

Figure 8–1. Fire Protection Equipment Inspection Report

Heating provisions should be checked for the dry pipe valves. Water tanks should be checked to determine if adequate protection against freezing is provided and that any heating systems are in good operating condition. The condition of fire pump reservoirs should be checked, as well as the suction intakes from water sources. Buildings should be surveyed to assure that cold air will not enter or expose sprinkler system piping to freezing.

Spring Inspection

As soon as the danger of freezing has passed in the spring months, an inspection should place attention on the reopening of cold-weather valves; testing, cleaning, and resetting dry pipe valves; testing water-flow alarms and conducting water-flow tests.

INSURANCE INSPECTIONS

Insurance carriers frequently pay special attention to sprinkler system reliability. Some insurance carriers offer sprinkler system testing services in the common interest of both the owner and the insurer. Through routine tests, sprinkler systems can be verified to be in good operating condition and any impairments can be revealed. Since such tests are made at the owner's responsibility and risk, intelligent cooperation in conducting the tests serves the best interest of the owner.

FIRE DEPARTMENT INSPECTIONS

Inspections of sprinkler systems are often made by many fire departments. These inspections verify that control valves are open and provide an opportunity to become familiar with water supply connections. Fire department inspections are customarily made by the fire company closest to the property or building. Owners and managers should utilize fire departments' services as a show of good faith to the public, the organization, and insurers.

CONTRACTORS' SERVICES

Standardized sprinkler system inspection and maintenance services are offered by sprinkler manufacturers and competent sprinkler contractors. These inspections are particularly advantageous to a property owner who must rely on an outside inspection service. These services can provide periodic examinations and reports. They are valuable to the property owner not only for monitoring the condition of sprinkler systems, but also because of valuable instruction that can be given to employees in the process. In addition to sprinkler systems, service often can be provided for other fire protection features such as water supplies and fire pumps (Davis, 1981). Inspection and maintenance services offered by sprinkler contractors normally follow a format that is acceptable to most insurance carriers.

CENTRAL STATION SUPERVISORY SERVICE

Central station supervision of sprinkler alarm and control devices provided under contract is an especially valuable aid to an organization. The outside party responsible for reporting to the owner or manager each incident involving water flow or gate closure or other supervised action keeps a constant check on the condition of the equipment and stimulates interest for the care of the overall system on the part of the plant fire organization.

RELIABILITY TESTS OF AUTOMATIC SPRINKLERS

Where sprinkler systems are subject to loading or corrosion, even moderately or slightly, they should be carefully and frequently examined. If the condition of a sprinkler system appears to be unreliable, then it should be removed from service. Parts and assemblies can be carefully packaged and sent for testing to Underwriters Laboratories, Inc., Factory Mutual Engineering Corporation, or the manufacturer. Care should be taken to minimize the period of interruption of protection and to make sure that all valves are left open after sprinklers are replaced (Planer, 1979). When installing sprinklers, or when removing sprinklers that are to be cleaned and reinstalled, special wrenches provided by manufacturers for their own size and shape of sprinklers should be used to prevent mechanical injury and distortion.

Accumulation of Foreign Material on Sprinklers

In many cases, conditions exist which cause an accumulation of foreign material on automatic sprinklers so that operation of the sprinkler may be retarded or prevented. This is commonly called *loading*.

Any accumulation of foreign material on sprinklers tends to retard their operation, owing to the heat-insulating effect of the loading material. If the deposit is hard, it may even prevent the sprinkler from operating. The best practice is to replace such loaded sprinklers with new sprinklers rather than to attempt to clean them. Attempts at cleaning, particularly where deposits are hardened, are likely to damage the sprinkler, rendering it inoperative or causing possible leakage.

Deposits of light dust, such as may be found on sprinklers in wood-working plants and grain elevators, are less serious than hard deposits. Dust may be expected to delay the operation of sprinklers, but ordinarily will not prevent the eventual discharge of water. Dust deposits can be blown or brushed off, but blowing by compressed air should not be undertaken where it can create a dust explosion or ignition hazard. If a brush is used, it should be soft to avoid possible damage to sprinkler parts.

Paint solvents, sometimes used in cleaning sprinklers, are not ordinarily injurious to solder or built-type sprinklers, but some solvents may damage sprinklers of the chemical type. The use of any flammable solvent for cleaning involves a hazard. Authorities advocate removing sprinklers and doing the cleaning outside of the building. This also permits a more thorough cleaning operation of immersing and rinsing the sprinklers in the solvent.

Cleaning, to be effective, must be thorough, as small quantities of paint or similar materials left between sprinkler parts, even though not conspicuous, may seriously delay or prevent sprinkler operation (Underdown, 1971).

Water-solution cleaning liquids containing caustic or acid components are likely to be injurious to sprinklers and should not be used for cleaning. No hot solution of any kind should be used for cleaning.

Sprinklers can be protected when ceilings or sprinkler piping are being painted by temporarily placing small, lightweight paper bags over them. Paper bags, however, are likely to delay the operation of the sprinklers, and they are to be removed immediately after the painting is completed. Sprinklers in spray booths present a special problem for which there is no satisfactory solution except to conduct the spraying process in such a manner that no spray will reach the sprinklers. If processes can be located to minimize deposits and cleaning is done frequently, conveniently accessible sprinklers may be cleaned without removing. Using a coating of grease or soft neutral soap facilitates washing or wiping off deposits. If grease is used, then it should be a grease with a low melting point. Unless cleaning is done carefully, deposits are likely to accumulate to such an extent as to seriously interfere with sprinkler operation. The use of paper bags to protect sprinklers in spray booths is a fairly common, but not a recommended, practice.

Corrosion of Automatic Sprinklers

Corrosive conditions are likely to make automatic sprinklers inoperative or retard the speed of their operation. Corrosive vapors may seriously affect not only the heat-actuated element and the valve-retaining members of an automatic sprinkler, but also may be severe enough to weaken or destroy other portions of the sprinkler. In most instances, such corrosive action is slow but sure, and thus must be vigilantly watched (NFPA, *Fire Protection Handbook*, 1997).

Some types of sprinklers are less susceptible than others to corrosive conditions. Non-ferrous metal is used for sprinkler parts, but special protective coatings are necessary for all types when exposed to extreme corrosive conditions. Approved corrosion-resistant or special, coated sprinklers are needed in locations where chemicals, moisture, or corrosive vapors exist. Representative occupancies having corrosive conditions likely to adversely affect sprinklers are given in *NFPA 13—Installation of Sprinkler Systems*. Corrosion-resistant Sprinnlor sprinklers fabricated with corrosion-resistant matter or with a special coating or platings may be used in an atmosphere that would normally corrode sprinklers.

Protection of Pipes Against External Corrosion

Under some conditions, corrosive vapors may cause rapid deterioration of steel pipe and hangers. This necessitates frequent replacement unless proper protection is provided. Under most conditions, cast-iron fittings will not be seriously affected.

There are two methods available for avoiding pipe corrosion: protective coatings and the use of materials other than steel. Under severe corrosive conditions, protective methods are not completely satisfactory. Genuine, wrought iron pipe or special-alloy, noncorrosive pipe will give the best results. Galvanized steel, under some conditions, may be the best and most economical method of obtaining reasonably long life for the piping system. This might apply to chemical plants, salt works, or similar properties where corrosion may be severe. Stainless steel and copper piping have also been used in some cases.

When corrosion of existing equipment becomes a maintenance problem, replacement or the application of a recognized type of protective coating are remedial measures. *NFPA 13* cautions that when a protective coating is applied to old piping, be sure first to remove all corrosion, scale, and grease. Otherwise, little benefit will be derived from the coating. Piping should be carefully examined at frequent intervals and if evidence of pitting, checking, blistering, or other failure is noted, the pipe should be cleaned and another coat of protective paint applied. A similar procedure is appropriate for sprinkler pipe hangers (Planer, 1979).

SPRINKLER SYSTEM IMPAIRMENTS

NFPA statistics and the records of large-loss building fires demonstrate the results of having impaired sprinkler systems when a fire occurs. Impairments arise if a sprinkler system water supply is shut off for extensions or alterations to the sprinkler piping, repairs due to accidental damage to piping or sprinklers, replacement of sprinklers after a fire, or maintenance or replacement of sprinklers and other sprinkler system devices. When sprinkler protection is interrupted, every effort must be made to limit the extent and duration of the interruption. A cardinal rule is to notify the fire department whenever such a sprinkler system impairment exists. This prevents the fire department from placing reliance on the systems. Most insurance companies also request that owners advise them when there is an interruption of sprinkler system protection. Insurance companies desire that alternate means of fire protection be arranged if the impairment is not temporary.

When sprinkler systems must be shut off, the work should be planned for a time when the least hazard exists. In industrial plants, about three times as many fires occur during operation as during idle periods. It is advisable that any sprinkler system repairs be done on a weekend or other idle period. Special watch services may be required to assure prompt detection of any fire which might develop while the sprinkler system is impaired. Sectional valves, rather than main valves, should be used where possible to take advantage of multiple water supplies.

All personnel, materials, and tools should be made ready before the sprinkler protection is impaired. If mains are to be opened, prepare wooden or other plugs or caps, and clamps to close the end of pipes quickly. Take emergency measures to maintain the maximum possible water supply to sprinklers. One possibility is to make a temporary hose connection from a domestic or industrial supply to the sprinkler riser. Adapters for connecting 2½-inch hose to sprinkler systems should be kept on hand. Procedures for supervising closed valves, for notifying the fire department and insurance companies, and for making water-flow tests after the work is completed should be followed.

BASIC PRINCIPLES OF MAINTENANCE AND INSPECTION

Care and maintenance of any fire protection feature assures optimum reliability. Major considerations for sprinkler system maintenance are (NFPA, *Fire Protection Handbook*, 17th edition):

- Sprinkler protection is complete in the areas protected.
- No obstructions exist that might inhibit the distribution of water discharge from the sprinklers.
- The water supply is constantly available.
- There is no opportunity for the sprinklers to freeze.
- All devices forming a part of a sprinkler system, alarm, supervisory systems, or water supply are in dependable operating condition.

Inspection is merely an organized, methodical procedure for visually determining the operating condition of components and assemblies. Inspectors should be qualified for the tasks they are performing. Inspections reveal the need for maintenance, repair, or remedy.

Inspection and maintenance functions are closely related and may overlap. Inspections frequently involve matters which may be classified as maintenance, and maintenance sometimes requires its own inspections and tests beyond those made routinely during so-called fire inspections. Management is responsible for their coordination.

INSPECTION AND MAINTENANCE OF FIRE EXTINGUISHERS

The term inspection means a visual check. This visual check determines that a fire extinguisher is ready to operate. An inspection also considers whether the extinguisher is fully charged and will function effectively when used for its intended purpose. *NFPA 10, Standard for Portable Fire Extinguishers*, details the inspection, testing, and maintenance requirements for various types of fire extinguishers.

An inspection should verify the following about a fire extinguisher (Davis, 1981):

- It is located at a designated place.
- It is conspicuously located.
- It is easily accessible.
- It is fully charged.
- It is untampered and has not been vandalized.
- It is protected from the environment and incidental damage.

The effectiveness of inspections depends greatly upon the frequency, regularity, and thoroughness with which they are done. Depending upon the size of the facility, it is recommended that the manager, property owner, or a designated person check the extinguishers at the beginning of each month. Where a plant fire brigade is organized, it is advisable for the fire brigade to be included in the inspection process in order to become familiar with the equipment and its location.

Maintenance, as distinguished from an inspection, requires that fire extinguishers undergo a thorough examination targeted at moving components and assemblies. Fire extinguisher maintenance should include:

- complete examination of each extinguisher
- any necessary repairs
- recharging
- replacement of any defective parts

Maintenance should be performed at least annually, after each use, or when an inspection reveals a problem. Similarly, if an inspection shows evidence of tampering, damage, or agent leakage, then a complete maintenance check should be initiated.

Fire extinguisher maintenance programs should include accurate record keeping. Fire extinguisher records should include the following information:

- The date of the maintenance
- The name of the person or organization performing the maintenance
- The date when last charged and the name of the person or organization recharging the extinguisher
- The hydrostatic test data and the name of the person or the organization conducting the hydrostatic test
- A description of dents or damage remaining after passing a hydrostatic test
- The date of the six-year maintenance for certain stored-pressure, dry chemical types

The persons who are responsible for performing maintenance procedures and record keeping should be appropriately trained. Fire extinguisher tags are commonly used as a convenient means for recording periodic inspections and maintenance. Generally, for routine inspections, a tag or pressure-sensitive label is used to record the date and the inspector's initials.

Seals and tamper detectors should be utilized in conjunction with an extinguisher inspection program. The seal or tamper indicator may consist of a thread, band, plastic insert, or other device that conforms to the standards of testing laboratories. Lead and wire seals were commonly used until plastic seals were introduced. As long as the device remains unbroken, there is reasonable assurance that the actuating mechanism of the extinguisher has not been used.

REFERENCES

Building Officials and Code Administrators, Inc. *BOCA National Building Codes*. Country Club Hills, IL: Building and Code Administrators, Inc., 1995.

Davis, Larry. "OSHA Standard For Fire Service." *Fire Engineering*, March 1981.

National Fire Protection Association. *Fire Protection Handbook*. 17th ed. Quincy, MA: National Fire Protection Association, 1994.

______. *Life Safety Code Handbook*. Quincy, MA: National Fire Protection Association, 1995.

______. *National Fire Codes*. Quincy, MA: National Fire Protection Association, 1995.

Planer, Robert. *Fire Loss Control*. New York: Marcel Dekker, 1979.

Underdown, G. W. *Practical Fire Precautions*. London: A. Wheaton and Company, 1971.

STUDY GUIDE QUESTIONS

1. What is the primary purpose of an automatic sprinkler system?
2. Who should maintain responsibility for assuring that adequate maintenance is performed?
3. How often should sprinkler system inspections be conducted?
4. What types of things should inspectors focus on when inspecting sprinkler systems?
5. When sprinklers are found to be "loaded" with foreign material, what actions should be taken for correction?
6. What are the best means for avoiding pipe corrosion?
7. What is one cardinal rule when repairing sprinkler systems that requires water to be shut off?
8. When repairing sprinklers in industrial plants, when should the repairs be scheduled?
9. What are some inspection services available other than self-inspection?
10. What are the four steps of extinguisher maintenance and how does it differ from inspection?

CASE STUDIES

1. A manufacturing facility has twelve sprinkler systems: ten wet pipe systems and two dry pipe systems. Develop an inspection, testing, and maintenance program for the sprinkler systems using applicable NFPA standards.
2. The same manufacturing facility has 150 multipurpose fire extinguishers. Develop a letter that requests bids from local companies for a 5-year fire extinguisher inspection and maintenance contract and includes some basic performance criteria.

Legal Aspects, Organization, and Legislation

CHAPTER 9

CHAPTER CONTENTS

LEGAL ASPECTS OF FIRE SAFETY

The legal aspects of fire safety are complex. Each management level not only has a moral responsibility to assure fire safety, but each also has legal responsibilities. Some of the major responsibilities and liabilities are described in this chapter. They are divided into these major areas:

- Upper-management responsibilities
- Safety management staff responsibilities
- Middle-management responsibilities
- Lower-management responsibilities
- Employee responsibilities

Each major area is discussed.

Upper-Management Responsibilities

Property management and fire safety can be profit-protecting activities for any organization. Failure to practice property conservation and fire prevention can and usually does produce a direct, tangible loss, which must be paid for with dollars that would otherwise be profit. Upper-management responsibilities include the following:

1. Making the policy statement: It is imperative that the policy statement be approved and issued over the signature of the top official in the organization. Without such a commitment the program will be continually compromised by opposing viewpoints of other managers.
2. Empowerment of Responsibility and Authority: The president or CEO empowers responsibility and authority to individuals of the management team, and appoints one individual to administrate the program. This person will report to top management.
3. Serving as Executive Officer: This person will be held accountable by federal agencies, state agencies, and the courts for serious violations of regulatory standards. Penalties can include fines and even criminal prosecution. This person has overall responsibility for the facility's safety and health program.

Safety Management Staff

The safety management staff should be responsible for providing input to top management regarding the policy, stimulating acceptance of sound safety practices, promoting an attitude of safety awareness, and working with top management to accomplish the objectives of the safety policy. It should be accountable for providing technical procedures, establishing safety procedures, issuing work instructions, and enforcing rules and standards in major divisions of the company. The fire prevention and safety department should be responsible for staff assistance to line and service departments in fire prevention, fire protection, accident prevention, and in the control of emergencies affecting the safety of persons or damage to buildings, equipment, and products. Some duties are outlined as follows:

1. To update knowledge of new developments in the field of accident and fire prevention
2. To prepare, review, and/or approve all applicable safe-practice standards
3. To review new or modified methods, materials, supplies, and equipment, including buildings machines, tools, and production devices
4. To develop a centralized program and to assist in departmental programs that promote and maintain interest in accident and fire prevention
5. To interpret laws, directives, and codes dealing with accident prevention and fire prevention
6. To investigate and report on safety suggestions or delegate such investigation to safety leaders
7. To accompany state and insurance inspectors on all inspection trips within the department

Middle-Management Responsibilities

The front-line manager maintains daily contact with production and maintenance workers. Employee safety and health and safe working conditions for a given area are vested in this person. The supervisor is the key person in the safety program because he or she is in constant contact with employees. The plant superintendent is expected to enforce compliance with the company safety policy by:

1. being familiar with the safety program and ensuring its effective application;
2. attending meetings of the plant safety committee and giving full support to all committee activities;
3. taking an active interest in the organization's safety training programs;
4. participating in investigations of all major and submajor injuries;
5. giving leadership and direction in the administration of safety activities;
6. conferring with the safety department when new operations are installed or new tools, equipment, and materials are introduced into the plant, to see that all proper safety precautions are taken for their safe use.

Middle management must support the establishment of a company fire brigade. The employees are trained in first aid, rescue, and fire fighting techniques in an effort to reduce losses from fires.

Lower-Management Responsibilities

The front-line foreman's job is divided into three parts: (1) to know the fire protective system and how it works in his department, (2) to cooperate in emergency assignments made to his or her department, (3) to see that those reporting to him or her work safely. The last means more than most supervisors realize. It means becoming property-conservation and fire-prevention-minded about everything in his or her area, making sure that everyone knows how to keep fires from starting; that everyone knows how to keep anything from happening to the fire protective equipment; and that nobody ever stops doing their share. Three other areas of concern to the front-line supervisor are (1) worker instruction; (2) protection of both workers and property; (3) fire: what to do before, during, and after.

Worker instruction consists of the following:

- conducting Job Safety Analysis and instructing employees in proper safety procedures for all work operations
- indoctrinating new employees in safety awareness
- training new employees in the proper job procedures

Protection of both workers and property includes:

- conducting housekeeping inspections
- reporting and/or correcting fire hazards
- conducting equipment inspections
- identifying and/or correcting hazardous conditions
- enforcing safety rules and regulations

Employee Responsibilities

Employees are expected to follow safe procedures and take an active part in protecting themselves, their fellow workers, and the plant. They should be encouraged to detect and report hazardous conditions, practices, and behavior in their workplaces and make suggestions for their correction. The ideal safe and efficient industrial operation is reached only when all employees are safety conscious. All employees should adhere to the following rules:

1. Comply with supervisor instructions.
2. Report all accidents and injuries immediately.
3. Submit recommendations for improving safety and efficiency.
4. Know your exact duties in case of fire or other catastrophe.

ORGANIZATIONAL STRUCTURE

Few realize the effect organizational structure can have in influencing behavior critical to the success of a loss control program. This is especially true in the workplace. This chapter will attempt to identify organizational concepts and principles which contribute to preferred results expected in a loss control program generally, and the fire plan specifically.

It must first be understood that the organization dictates the organizational structure within which the safety department, or any other department, will operate. Thus, to operate successfully, the safety manager must learn to function within the confines of the imposed structure, both formal and informal. It can generally be assumed that in fulfilling its mission, the safety department will function with less power and authority than it would like.

Therefore, it is imperative that the safety practitioner understand the various concepts of authority. Authority is the right to decide or act. Line authority is the simplest type of authority, with each position having direct and general authority for taking actions and complete authority over lower positions in the hierarchy. Staff authority is purely advisory to the position of line authority. The individual with staff authority studies a problem, seeks alternatives, makes recommendations, but has no authority or power to require that the recommendations be put into action. Functional authority focuses on achieving the advantages of specialization by permitting the staff specialist to enforce his orders within a limited and well-defined scope of authority.

The line manager with appropriate authority is the backbone of the organization, but staff positions provide the supplemental advice and direction which allow the line generalist to function successfully. The safety practitioner might actually serve in a staff position with pertinent authority for some programs and functional authority for others. A failure of the safety manager to know and understand his or her authority in various situations could result in serious conflicts and substantially hinder the success of his or her program.

Although it would be quite desirable for the safety manager to have functional authority over all his programs, it is quite likely to expect staff authority in many cases. As a result of this limited type of authority, success for the program will depend on gaining a collaborative effort of line managers and staff personnel in order to identify and correct safety problems.

As noted above, functional authority is the preferred means, but this also may introduce organizational problems by breaking the Unity of Command Principle.

The Principle of Unity of Command requires that each individual report directly to only one boss. When the safety manager is given functional authority, his orders to individual workers may be in conflict with those of the workers' line supervisor. Naturally, it will be wise for the safety practitioner to exercise authority with care, so that he or she does not alienate the workers whose support is needed to carry out the safety program.

In the Total Quality Management paradigm, the manager's function changes from director, controller, and inspector to coach, facilitator, and team manager. Employees exhibit competencies in self-direction, self-development, and team-development skills. The Total Quality Management paradigm imbues the employees with a risk-taking spirit. Staff positions provide the input to create mutually agreed upon decisions which allow the organization to function successfully. Success depends ultimately upon the ability to fill management positions with people whom other people respect. Success in fire safety management depends on an effort of managers and personnel to identify and correct fire hazards.

The final organizational concern deals with informal organization and communication. In many organizational structures it is necessary to cross over formal lines of communication to effectively accomplish a task. It is quite clear that the broad scope of the loss control program will require such actions.

Even though the development of fire safety procedures is the responsibility of those having experience in fire protection, all personnel have responsibility for the program's execution. Upper management has the authority to administer the program by writing objectives establishing each department's responsibilities. The safety staff is responsible for assisting each department with fire prevention, fire protection, and emergency control. Middle management participates in program activities, assuring compliance with standards and the organizing of fire brigades.

Line supervisors must be trained in the use and maintenance of fire protection systems. They are responsible for the cooperation of their workers in emergencies. When employees are trained in their responsibility and encouraged to provide input to the fire safety management program, hazards are minimized and losses reduced. Employees comply with supervisors' instructions, report incidents and injuries, make recommendations, and perform their assigned duties during fires.

Despite a fire safety management program, losses cannot be minimized if managers and employees do not participate actively. The insurance industry insists upon the continuous participation of personnel in loss prevention, inspections, and the implementing of emergency procedures, as conditions of coverage. Insurance underwriters frequently accept organizations with major fire protection problems if there is evidence of effective guidelines for personnel in an organization's program, but they are unlikely to underwrite an organization which has given little consideration to the role of personnel in developing its programs.

Well-managed organizations recognize that most disastrous fires are the result of human error. Discrepancies are due to management's failure to organize and instruct its employees. By ensuring the participation of all personnel in fire safety management, benefit exceeds cost. A comprehensive fire and emergency response plan aids in reducing losses due to fire, explosion, windstorms, water damage, earthquakes, riots, or bomb threats. The plan should include emergency procedures, evacuation routes, and a listing of emergency services. It should be made available to all personnel.

FEDERAL LEGISLATION, AGENCIES, AND REGULATIONS

Safety professionals must be aware of applicable federal regulations in the respective geographical area in which the safety professional is operating. All legislation is aimed at protecting life, society, and property. The safety professional should be familiar with all legal aspects of legislation.

Federal Fire Prevention Control Act of 1974

The Federal Fire Prevention and Control Act of 1974 was the first congressional legislation attempting to regulate fire prevention activities. The key points of the act are as follows:

1. To educate the public about fire and fire prevention
2. To develop a program for fire fighting technology
3. To establish a National Academy of Fire Prevention and Control
4. To establish a National Fire Data Center
5. To establish a program that encourages state and local governments to develop master plans for fire prevention and control
6. To initiate a system to review and revise state and local fire and building codes
7. To organize and participate in a National Conference on Fire Control

The problem of fire has always been a major concern for our nation. On May 4, 1973, the National Commission on Fire Prevention and Control submitted a report of a two-year study on the loss of property and life by fire in the United States. This report identified the many problems of fire and its control, and recommended a continued federal focus upon the subject. Consequently, in October 1974 President Ford signed into law the Federal Fire Prevention and Control Act, establishing the United States Fire Administration as a functional operating unit of the Department of Commerce with the purpose of addressing the needs and major concerns of America's fire problem.

In 1978, under President Carter, the Fire Administration was consolidated along with the Federal Disaster Assistance Administration and the Federal Preparedness Agency, as well as earthquake, antiterrorism, and the emergency broadcast programs, into a single structure called the Federal Emergency Management Agency (FEMA). FEMA was then established as an independent agency of the Executive Branch with the purpose of administering disaster mitigation and response programs. As a result of this action, the Fire Administration is now able to work closely with the other units of FEMA on the many areas of emergency prevention, response, and recovery.

The Fire Administration recognizes that fire remains a state and local problem. The purpose, then, of the National Fire Administration is to support and reinforce state and local efforts, as well as the national effort for fire prevention and control. Some of the needs that this administrative agency focuses upon are:

- A uniform, broad-based pool of data with which to identify American's fire problem
- Safer homes through education and technology
- Protecting fire fighters from death and injury
- Comprehensive fire prevention and control planning at all levels of government
- Conquering arson
- Improved education and training for the nation's fire protection community
- A basic understanding of fire and its effects
- Providing assistance to state and local governments
- Providing a focus for the federal fire community

In order to meet these needs and to improve the effectiveness of state and local efforts, intensive work has been done to identify the priorities at the state and local levels; to develop new and improved fire prevention and control techniques; to test those techniques; and to provide the leadership, incentives, and methods to implement those techniques.

Organizationally, the United States Fire Administration (USFA), an office of the Federal Emergency Management Agency (FEMA), has been divided into four operating units:

1. Office of Fire Policy and Coordination
2. Office of Fire Fighter Health and Safety
3. Office of Fire Prevention and Arson Control
4. Office of Fire Data and Analysis

In addition, the United States Fire Administration is headed by the Office of Administrator who is in turn serviced by an Office of Chief Counsel, an Office of Administration, and an Office of Information Services. Another agency, the National Bureau of Standard's Center for Fire Research which is under the administrative jurisdiction of the Assistant Secretary for Science and Technology, closely coordinates and interlocks its programs with those of the USFA.

OSHA Act of 1970

Established as part of the Department of Labor in 1970, the Occupational Safety and Health Administration (OSHA) is responsible for establishing and enforcing workplace safety and health standards. It was created to assure every working man and woman in the nation safe and healthful working conditions. Subpart L of the OSHA Safety and Health Standards (29 CFR 1910) is primarily concerned with establishing requirements and standards for fire brigades, all portable and fixed fire suppression systems, fire detection systems, and fire- or employee-activated alarm systems installed to meet the fire protection requirement of this OSHA manual. These requirements apply to all areas of employment except for those of maritime, construction, and agriculture.

Respiratory Protection Standard

The new standard, which became effective April 8, 1998, replaces the respiratory protection standards adopted by OSHA in 1971 (29 CFR 1910.134 and 29 CFR 1926.103). It applies to general industry, construction, shipyard, longshoring, and marine terminal workplaces, but excludes agriculture. The standard requires employers to establish or maintain a respiratory protection program to protect their respirator-wearing employees. The standard contains requirements for program administration, including worksite-specific procedures; respirator selection; employee training; fit testing; medical evaluation; respirator use; respirator cleaning, maintenance, and repair; and other provisions. The standard also simplifies respirator requirements for employers by deleting respiratory provisions in other OSHA health standards that duplicate those in this standard, and revising other respirator-related provisions to make them consistent. In addition, the standard addresses the use of respirators in Immediately Dangerous to Life or Health (IDLH) atmospheres, including interior structural fire fighting. During interior structural fire fighting (an IDLH atmosphere, by definition), self-contained breathing apparatus is required and two fire fighters must be on standby to provide assistance or perform rescue when two fire fighters are inside the burning building.

The federal standard will apply only to federal employees and to private-sector employees who are involved in fighting fires. Federal OSHA has no jurisdiction over the many fire fighters who are state and local government employees or volunteers. Although OSHA has no jurisdiction over public-sector fire fighters, the 25 states operating OSHA-approved state plans do cover those workers. The states that have their own plans are expected to adopt revised respiratory protection standards. These plans may differ, but must provide protection to workers equivalent to the OSHA standard. It is through these state plan standards that the "two-in/two-out" requirements will be applicable to fire fighters in those states.

Based on the record in this rulemaking and the Agency's own experience in enforcing its prior respiratory protection standards, OSHA has concluded that compliance with the standard will assist employers in protecting the health of employees exposed in the course of their work to airborne contaminants, physical hazards, and biological agents, and that the standard is therefore necessary and appropriate. The respiratory protection standard covers an estimated 5 million respirator wearers working in an estimated 1.3 million workplaces in the covered sectors. OSHA's benefit analysis predicts that the standard will prevent many deaths and illnesses among respirator-wearing employees every year by protecting them from exposure to acute and chronic health hazards. OSHA estimated that compliance with this standard will avert hundreds of deaths and thousands of illnesses annually. The annual costs of the standard are estimated to be $111 million, or an average of $22 per covered employee per year.

Federal Mine Safety and Health Act

The 1969 Coal Mine Health and Safety Act, which was later amended by the Federal Mine Safety and Health Act of 1977 (P.L. No. 95-164) provides the regulatory body for mining. This act, which also repealed the Metal and Nonmetallic Mine Act of 1966, allowed for the regulation of the mines to be done under one system of regulations instead of the many laws and regulations which had previously governed mine operations.

The Act also transferred the responsibility of mine safety from the Department of the Interior to the Department of Labor. The Department of Labor then formed a new agency to regulate the mines called the Mine Safety and Health Administration (MSHA). Also established was an independent Federal Mine Safety and Health Review Commission to hear challenges to MSHA citations.

MSHA has jurisdiction over any work activity that goes on in a mine or on mine property. This includes the roads leading to the mines, the roads located on mine property, and all other mine-site operations, such as coal-preparation facilities.

STATE AGENCIES AND REGULATIONS

State Fire Marshal

State fire marshal offices were first established shortly before the turn of the century. Presently, there are 49 active state fire marshals. In many areas of the country, the fire marshal is assisted by deputy and/or assistant fire marshals or their equivalents. Approximately 50% of these offices are funded by taxing fire insurance premiums. Others receive their funding from various fees such as explosive permits, electrician license fees, and inspection and plan review fees.

Fire marshals are responsible for enforcing applicable fire safety laws. Such fire safety laws might include provisions for the following:

- prevention of fires by enforcing compliance with state fire codes;
- storage, sale, and use of combustibles and explosives;
- installation and maintenance of automatic fire protection systems;
- assessment of means and adequacy of exits on all state properties and in public places;
- investigation of the cause and origin of fires;
- presentation of fire prevention education;
- review of plans and specifications for new construction, alterations, renovations, and additions;
- conducting of fire drills in the public school system.

The Office of the Attorney General is responsible in some states for providing legal assistance in the enforcement of fire prevention codes. In other states, the state police directly operate a fire marshal's office as part of their locally designated duties.

State fire marshal offices may be the primary code enforcement agency in some states. In other states, an assistant fire marshal, locally designated fire marshal, or municipal fire chief may serve as an ex-officio deputy. In these cases, the state office acts as a back-up agency, rendering assistance, support, and expertise in difficult situations.

State Insurance Commission

This agency, or its equivalent, is generally present in most states. It is responsible for determining fair insurance rates for various municipalities within a state. Accordingly, to qualify for lower rates, more codes and laws must be enforced. Once a certain level of proficiency is met by a municipal fire department, local governments expect continued compliance to preclude any additional insurance levies. This office may be considered an indirect means of enforcement of state codes.

County and Municipal Ordinances and Codes

The enforcement of county and local codes and ordinances affecting fire safety is frequently divided among different agencies and is organized in different ways in different areas. Generally, these codes and their enforcement are handled at the municipal level to avoid duplication of effort. The codes usually consist of adaptations of national codes, state laws, and municipal regulations.

THE STRENGTH OF LAWS

For the most part, limited police powers only permit control of matters in the general public's interest, as distinguished from the interest of the individual. The scope of municipal fire legislation is limited to that which is considered reasonable, and the general public interest is balanced against that of the individual wherever there may be conflict, particularly where fire protection in the public interest may call for expenditures beyond those which a property owner considers essential to his or her own interests. In such matters, codes and ordinances generally favor the safety and property of the public at large.

The number of inspectors can vary with the size of municipality and the amount of revenue allocated toward an effective inspection program. These factors have a direct influence on how well codes are enforced. Limited manpower and resources severely impair an effective inspection program.

Building Department Enforcement of Building Codes

All construction requirements are part of municipalities' building codes, including requirements for exits and for fire extinguishing equipment, but usually maintenance of these items is covered by the fire prevention code. Enforcement is normally the responsibility of the fire marshal, building inspector, or the Fire Prevention Board. In some instances, the inspector or marshal at the county level may be appointed by the state fire marshal. Variances and appeals are considered by the Fire Prevention Board. In some counties and municipalities, the prosecuting attorney or the district attorney is empowered to act as the fire marshal's representative in prosecuting violators of fire codes and ordinances.

LEGAL RIGHTS OF FIRE DEPARTMENTS

The legal rights of fire departments are specified at various levels of government. They are briefly stated within the National Fire Codes, written into state laws, and are further defined in city ordinances and charters. The following rights represent an overview of the legal rights of fire fighters in most states.

Right of Entry

Right of entry is restricted to two basic occasions:

1. Fires in Progress. Fire fighters shall have the right to enter any premises at any time when a fire is in progress, where there is reasonable cause to believe a fire is in progress, or for the purpose of extinguishing a fire.

2. Premises Protection. Fire fighters shall have the right to enter any premises adjacent thereto for the purpose of protecting such adjacent building or premises, or for the purpose of extinguishing the fire which is in progress in an adjacent building or premises.

Authority When Answering an Alarm

1. All bystanders and other persons shall obey all proper orders duly given by the chief, fire officer, and/or the subordinates at a fire.
2. The chief or designated commanding fire officer will have the authority to maintain order at the fire or its vicinity and direct the actions of the firemen at the fire.
3. Keep bystanders or other persons at a safe distance from the fire and fire equipment.
4. Facilitate the speedy movement and operation of fire fighting equipment and firemen.
5. Until the arrival of a police officer, direct and control traffic and facilitate movement of traffic.
6. The chief or commanding fire officer shall display his firefighter's badge or proper means of identification.
7. Authority granted extends to the activation of traffic control signals designed to facilitate the safe egress and ingress of fire fighting equipment at a fire station.

Taking and Preserving Property

The chief or fire officer in command is authorized and empowered to take and preserve any property which indicates that the fire was intentionally set.

Conducting Investigations to Determine Cause of Fire

1. The chief or other designated fire officer may enter the scene of such a fire within a reasonable time after the fire has been extinguished (U.S. Supreme Court decision).
2. If there is evidence that a fire was of incendiary origin, authorized fire fighting personnel may control who enters such scene by posting no-trespassing signs at the fire scene after such fire has been extinguished.
3. In order for said owner, lessee, or person to recover or salvage personal property from a posted scene, first, it must be declared safe by authorized fire department or company officials; he or she must be accompanied by, or granted permission to enter the scene by, an authorized fire department or company official.

Attacking, Hindering, or Obstructing Firemen or Emergency Equipment

It shall be unlawful, while any fire department or company or firefighter is in the process of answering an alarm of fire, or extinguishing a fire, or returning to station, for any person to:

1. attack any firefighter or fire fighting equipment or emergency vehicles with any firearms, knives, fire bombs, or any object endangering life or property;

2. take any action for the purpose of hindering or obstructing any firefighter, equipment, or emergency vehicle by any means; or
3. refuse to take any action for the purpose of hindering or obstructing any firefighter, equipment, or emergency vehicle by any person.

REFERENCES

Bird, Frank E., and Germain, George L. *Practical Loss Control Leadership*. Loganville, GA: International Loss Control Institute, Inc., 1992.

Blackburn, Richard, and Rosen, Benjamin. "Total Quality and Human Resources Management: Lessons Learned from Baldridge Award Winning Companies." *Academy of Management Executive* 7 (3) (1993):49–66.

Building Officials and Code Administrators, Inc. *BOCA National Building Codes*. Country Club Hills, IL: Building Officials and Code Administrators, Inc., 1995.

Clet, Vince H. *Fire-Related Codes, Laws and Ordinance*. Beverly Hills, CA: Glencoe Press, 1978.

National Fire Protection Association. *Fire Protection Handbook*. 17th ed. Quincy, MA: National Fire Protection Association, 1991.

______. *Life Safety Code Handbook*. Quincy, MA: National Fire Protection Association, 1995.

______. *National Fire Codes*. Quincy, MA: National Fire Protection Association, 1995.

Robertson, James C., and Benseger, Bruce. *Introduction to Fire Prevention*. London, England: Glencoe Press, 1975.

United States Supreme Court, *Michigan v. Tyler*, May 31, 1978.

U.S. Commission of Fire Prevention. *America Burning*. Washington, D.C.: U.S. Government Printing Office, 1973.

Workplace Safety in Action: Hazard Assessment. Neenah, WI: J. J. Keller and Associates, 1993.

Emergency Response Planning for Safety Professionals

CHAPTER 10

CHAPTER CONTENTS

INTRODUCTION

In the wake of the September 11, 2001 attack on the New York City World Trade Center, the astonishingly low casualty count is a tribute to how far into our culture emergency response initiatives have penetrated. Terrorists, both international and "home grown," have demonstrated that they have the knowledge and capabilities to strike anywhere in the world. Realizing this, safety officials of at least one major tenant in the World Trade Center had identified the possibility of the building being hit by a fuel-laden plane, and developed effective, rapid evacuation procedures.

History has shown that no community is immune to any emergency situations, that disasters transcend all geographic and demographic boundaries.

The daily problems we face are now more complex than ever and much different from those we faced a generation ago. Environmental changes, economic growth, technological advances, and new threats in this country and around the globe have created challenges for our society and the safety profession.

FEDERAL EMERGENCY MANAGEMENT

According to the Federal Emergency Management Administration (FEMA), all jurisdictions—suburban, urban, and rural—are potential areas for some type of emergency situation. In the year 2000, natural disasters alone claimed the lives of more than 50,000 people around the world and resulted in economic losses exceeding 90 billion dollars, according to Kay Goss, FEMA's Associate Director for Preparedness and Training.

People generally do not *expect* an emergency or disaster to occur, especially one that affects them or their fellow employees. We must recognize that people always need to be aware of their surroundings and have a duty to know what they should do if an emergency arises. The protection of people and property is the most important thing an individual can accomplish during any emergency.

There are a number of approaches that have been developed by FEMA to bring together safety professionals in industry, universities, hospitals, schools, and other venues to discuss the needs of emergency management professionals and program development, as well as general items of interest pertaining to hazards, disasters, and emergency management problems.

WORKPLACE EMERGENCIES

It is important for each business to design an action plan to ensure the safety and protection of employees and property in the event of a disaster. The protection of employees, enterprise assets, and the community in which the company is located are critical to any corporation. This action response plan will provide established guidelines, policies, and procedures for safety professionals to follow when faced with an emergency.

The implementation of a comprehensive disaster response plan should be established through the company's management team, directed by senior management in concert with the company fire brigade. This approach will determine possible solutions to emergency-related problems and evoke recommendations that can improve the readiness of the workers in the facility.

KEY ELEMENTS OF THE EMERGENCY RESPONSE PLAN

The emergency response action plan should be designed to accomplish the following objectives:

- Improve safety awareness and emergency/disaster readiness.
- Protect the lives and assets of the corporation.
- Assign specific emergency responsibilities to employees relative to their competencies and normal work functions. This action begins with the fire brigade, whether it is full time or part time.
- Provide training and preparation for fire brigades and employees.
- Provide for orderly and efficient transition from normal to emergency procedures.
- Review the plan with the local Office of Emergency Services (OES) and include key telephone numbers.
- Develop a crisis communication plan for dealing with the media.
- Reduce losses associated with emergencies and disasters through improved corporate resources.
- Purchase and maintain equipment and supplies necessary for emergency situations.
- Provide the communication and transportation systems needed during potential problem situations.
- Establish site evacuation routes and procedures, both primary and secondary.
- Have safety professionals advise and lead during simulated drills based on the emergency action plan.
- Periodically evaluate and revise the emergency response plan.
- Document evaluation results and corrective actions and incorporate them into a revised plan.

TYPES OF EMERGENCIES—NATURAL OR MANMADE

A workplace emergency is an unforeseen situation that threatens employees, customers, or the general public; disrupts or closes down the business; and can cause physical or environmental damage. Emergencies may be natural or manmade and may be precipitated by the following:

- Hurricanes
- Tornadoes

- Floods
- Fires
- Civil disturbances
- Terrorism
- Chemical spills
- Toxic gas releases
- Explosions
- Radiological accidents
- Workplace violence resulting in bodily harm or trauma
- Sabotage

Leadership and direction are key components in all emergency response programs. It is essential that industry and business emergency response plans be properly prepared to handle such conceivable incidents, keeping them in the realm of emergency rather than allowing them to escalate into a disaster.

ALERTING AND WARNING EMPLOYEES

The emergency response plan must include guidelines for workers on how to report emergencies, as required. It must also include a way to alert employees, including disabled workers, to evacuate or take other necessary actions. All aspects of existing warning systems must be identified, and provisions may be made to implement all of them as needed.

Safety managers must receive timely information on possible threats to the workplace site and be able to transmit that information to key staff members and all other employees. Among the steps for alerting employees are the following:

1. Make sure alarms are distinctive and recognized by all workers as a signal to evacuate the workplace or perform actions identified in the response plan.
2. Inform all employees of the warning system that will be used to alert them to danger. Provide an alternate means of warning that will back up the primary system.
3. Define the responsibilities of departments or personnel and describe activation procedures.
4. Secure an auxiliary power supply in the event that electricity is turned off.
5. Refer to 29 CFR 1910.165 (b)(2) for more information on alarms.

Although not required by OSHA, safety professionals may also want to consider the following:

1. Various devices to alert employees not able to recognize an audible or visual alarm.
2. An updated list of key personnel, based on priorities (plant manager, physician), to notify in the event of an emergency during off-duty hours.

ACCOUNTABILITY AFTER EVACUATION

It is critical to account for all employees following an evacuation. Problems in the assembly area can lead to delays in the rescue operations for people trapped in a building. The fastest and most accurate way to account for employees is to incorporate and follow the steps below as part of the emergency response plan.

1. After evacuation the employees should gather in a predesignated assembly area.
2. A roll call should be administered in the assembly area. The names and last known location of anyone not accounted for should be submitted to the official in charge.
3. A method should be established to account for nonemployees, such as customers and suppliers.
4. In case the incident worsens, procedures for further evacuation should be established. This may consist of sending employees home by normal means or providing them with transportation to an off-site location.

Management should include employees in the development of the emergency response plan. Encourage the employees to offer suggestions about potential hazards, worst-case scenarios, and proper emergency responses. After developing the plan, review it with your employees to make sure everyone knows what to do before, during, and after an emergency. Make sure that all employees receive proper training for emergencies. Hold staff meetings periodically to review the procedures, and keep a copy of the emergency response plan in an easily accessible location.

TRAINING EMPLOYEES ON TYPES OF EMERGENCIES

Train all employees relative to the type of emergencies that may occur and the course of action to be taken. Be sure employees understand the function and elements of the emergency response plan, including types of potential emergencies, reporting procedures, alarm systems, evacuation plans, and shutdown procedures. The fire brigade should discuss any special hazards that are on site such as flammable materials, toxic chemicals, radioactive sources, or water-reactive substances. Brigade leaders should clearly communicate with all employees in leadership roles during an emergency to minimize confusion.

Key Points for Training Employees

Employee training is crucial for most activities within any company and especially when it deals with emergency response. Proper training in this area will prove an important asset to the entire enterprise, as well as to each employee, if and when it is needed.

Specific training may involve brigade team leaders as well as emergency responders from the outside. Training should include the use of fire extinguishers, power disconnects, emergency response procedures, search and rescue techniques, first aid, and medical treatment such as CPR.

General training should include the following:

- Emergency response procedures
- Individual roles and responsibilities
- Hazards, threats, and protective procedures
- Notification, warning, and communication procedures
- Location and use of common emergency equipment
- Shelter and accountability procedures
- Facility shutdown procedures
- Means of locating family members in an emergency

The corporation should train its employees in first-aid procedures and respiratory protection, including use of an escape-only respirator; protection against blood-borne pathogens; and methods for preventing unauthorized access to the site.

Once the emergency action plan has been reviewed with all employees and everyone has had the proper training, it is a good idea to hold simulated drills. Drills should be scheduled as often as necessary to keep employees prepared, but at least once a year. A critique should be held after each drill with management and employees to evaluate the effectiveness of the drill. Once the strengths and weaknesses of the plan have been identified, management should be encouraged to improve it with additional elements.

CONTINUITY OF MANAGEMENT

Specific measures should be developed to ensure continuity of leadership during any emergency. These should include:

- Maintaining a continuous chain of command.
- Developing and establishing lines of succession for key officers and operating personnel.
- Providing for an alternate corporate headquarters.

Preservation and protection of vital records in an emergency is essential for quick return to normal operations. The vital-information protection program is an administrative tool for safeguarding records. Management starts by systematically determining what information is vital and which records contain this information.

SUMMARY

Key to implementing a successful emergency response action program is to make sure that it is comprehensive. The plan will provide established procedures and guidelines for management and staff to follow in the event of an emergency that will improve the readiness of the workers and their facility. The plan should be designed and developed to ensure the safety and well-being of all employees and to protect property at the workplace site.

REFERENCES

Anton, Thomas J. *Occupational Safety and Health Management*. Boston: Irwin, McGraw-Hill, 1989.

Bass, Lewis. *Product Liability: Design and Manufacturing Defects*. Colorado Springs, CO: McGraw-Hill, 1986.

Campbell, R. L., and Langford, R. F. *Fundamentals of Hazardous Material Incidents*. Chelsea, MI: Lewis Publishers, 1991.

Christoffel, Tom, and Gallagher, Susan. *Injury Prevention and Public Health*. Gaithersburg, MD: Aspen Publishers, Inc., 1999.

Della-Giustina, D. E. *Planning for School Emergencies*. Reston, VA: School and Community Safety Society of America, 1998.

Federal Emergency Management Administration. *Emergency Management Curriculum and Training—FEMA*. Emmitsburg, MD: Federal Emergency Management Administration, April 1999.

Hoffner, R. E. "EMI Training—Well Worth the Effort," *Emergency Management IAEM Bulletin*, May 2001.

National Safety Council. *Accident Prevention Manual for Industrial Operations*, 12th ed. Itasca, IL: National Safety Council, 2000.

______. *Injury Facts*, 2001 edition. Itasca, IL: National Safety Council, 2001.

Occupational Safety and Health Administration (OSHA). *Hospital and Community Emergency Response—"What you need to know."* Washington, D.C.: U.S. Department of Labor, 1997.

Purpura, Philip. *Security and Loss Prevention*, 3rd ed. Boston: Butterworth-Heinemann, 1998.

Schneid, Thomas D., and Collins, L. *Disaster Management and Preparedness*. Boca Raton, FL: Lewis Publishers, 2001.

STUDY GUIDE QUESTIONS

True and False

1. True False In the year 2000, natural disasters claimed the lives of 200,000 people around the world.
2. True False A key element within the emergency response action plan is providing a way to protect the lives of employees and assets of the corporation.
3. True False A workplace site emergency is caused only by natural disaster.
4. True False Continuity of management should provide for a continuous chain of command.
5. True False A critique should be implemented at the start of the emergency action plan.
6. True False Preservation and protection of vital records in an emergency is essential for a quick return to normal operations.

7. True False Terrorism does not constitute a workplace emergency.
8. True False According to FEMA, all jurisdictions—suburban, urban and rural areas—could experience some type of emergency situation.
9. True False Fire brigades should not discuss any special hazards on site within the workplace
10. True False Various devices to alert employees not able to recognize an audible or visual alarm should be part of the alerting and warning program.

CASE STUDY

Power Plant Explosion

The following case study depicts the description of the event and concludes with the summary and conclusions. This model would encourage a group discussion by enlarging on the different levels of performance in discussing a power plant explosion. Names and location of this incident have been changed to give a completely different appearance.

Date of Incident:	November 13, 1999
Time of Incident:	1:00 PM
Location of Incident:	Cheshire Power Station, Cheshire MA
Losses Incurred:	Severe plant damage, lost-time injuries, power generation

Plant Description

The Cheshire Power Station, located just upriver from Pittsfield, Massachusetts, has two 550-megawatt electric-generating coal fed boilers. The plant was constructed in the early 1970s and is a part of the State Power Systems Utility. The two units are Combustion Engineering Boilers that are fired on pulverized coal crushed in CE-Raymond Roller Mills located inside the plant building. Normal operating conditions of the boilers are 1000 degrees and 3600 psi at full load.

Description of the Event

On November 12, 1999 there was an indication of a fire in the 1A coal pulverizer. The supervisor in charge made the decision to spray water on the outside jacket of the pulverizer in an attempt to cool the hotspot. The water appeared to have cooled the hotspot and everything appeared to return to normal operations when temperatures in the mill dropped from 150 degrees to 130 degrees. The day shift relieved the night shift on the morning of November 13 and continued to observe the 1A mill with no indications of any trouble.

At approximately 1:00 PM a series of explosions occurred in and around the 1A pulverizer mill. The initial explosion separated a discharge pipe leading from the pulverizing mill to the boiler, releasing coal dust into the atmosphere. The explosion stirred the coal dust in the atmosphere, creating a rolling fireball that ripped through the plant in all directions, feeding on the fugitive plant structure. The force of the explosion downed several internal block walls near the 1A mill. Numerous small fires were started throughout the plant in cable trays and any combustible material. Several employees were injured with burns ranging from mild to very severe. The sprinkler system throughout the plant operated as designed and dept damage from fires to a minimum. The sprinklers also kept the flammable materials from igniting, including a large storage of hydrogen used for turbine cooling.

Post-Response Assessment

Analysis and investigation by the Cheshire plant management and engineering staff concluded that several factors contributed to the explosion of the CE-Raymond Roller Mill. The analysis also pinpointed several conclusions for improving the system to ensure explosion prevention during future operation.

- The design of the mill included a fender located in the grinding section of the mill that allowed for the buildup of material in several locations. The mill should be redesigned to stop the buildup of materials that could cause a hazardous situation.
- The associated pulverized coal piping system was coupled together to allow for expansion and contraction during load swings, creating a weakness in the system and a place for escape in the event of overpressurization. A stronger, more rigid system should be designed to prevent escape.
- Pulverizers should be redesigned to include a water supply and steam inerting system in the event of a fire.
- A contiuous gas monitoring system should be installed that warns of explosive gas buildup before an ignitable atmosphere is present.
- Improvements in policy, procedure, and training should be made for mill-fire and coal-fire situations.

Summary and Conclusions

In summary, the fire protection sprinkler system helped to prevent additional fire damage following the mill explosion. This protection is consistent with *NFPA National Fire Code 850, Electric Generating Plants*; more specifically, *NFPA 8503, Pulverized Fuel Systems.* Quick action by the on-shift operations support staff in initiating first aid and emergency response helped to save the lives of the severely burned employees. Changes made in attitudes, training, equipment, and procedures since the incident have made the process much safer.

REFERENCES

Explosion Incident Report. Cheshire, MA: Cheshire Staff, 1999.

National Fire Protection Association. *National Fire Codes*. Quincy, MA: National Fire Protection Association, 1994.

The United States Fire Administration

CHAPTER 11

CHAPTER CONTENTS

THE FIRE ADMINISTRATION'S MISSION

The mission of the United States Fire Administration (USFA) is to limit the loss of lives and economic assets caused by fire and related emergencies through leadership, advocacy, coordination, and support. USFA serves the nation independently, in coordination with federal agencies, and in partnership with fire protection and emergency services in communities. With a commitment to excellence, USFA provides public education, training, technology, and data initiatives.

The National Fire Problem

Year	Fires	Deaths	Injuries	Direct Dollar Loss (in millions)
1990	2,019,000	5,195	28,600	9,385
1991	2,041,500	4,465	29,375	10,906
1992	1,964,500	4,730	28,700	9,276
1993	1,952,500	4,635	30,475	9,279
1994	2,054,500	4,275	27,250	8,630
1995	1,965,500	4,585	25,775	9,182
1996	1,975,000	4,990	25,550	9,406
1997	1,795,000	4,050	23,750	8,525
1998	1,755,500	4,035	23,100	8,629
1999	1,823,000	3,570	21,875	10,024

Twenty-six years ago the passage of the Federal Fire Prevention and Control Act (Public Law 93-498) created a federal agency to focus on the nation's fire problem. Originally called the National Fire Prevention and Control Administration, it was later renamed the United States Fire Administration.

USFA REORGANIZATION

The reorganization structure for the U.S. Fire Administration was thoroughly discussed and well thought out. It was a process that involved representation from the bargaining unit and management alike at a level that was unprecedented in the history of the Federal Emergency Management Administration. Currently, the management team provides direction and leadership to move the organization forward, and the director has honored his commitment by enhancing the budget and supporting the new organizational structure.

USFA's 5-year operational objectives, beginning in 2002, were:

1. Reduce the loss of life from fire by 15 percent.
    - Achieve a 25 percent reduction in the age group of infants to 14-year-olds.
    - Achieve a 25 percent reduction among those 65 and older.
    - Achieve a 25 percent reduction among fire fighters.

2. See that 2500 communities have a comprehensive multi-hazard risk reduction plan in place, led by or including the local fire service.
3. Respond appropriately in a timely manner to any emergent issues.

A BRIEF HISTORY OF THE UNITED STATES FIRE ADMINISTRATION AND THE NATIONAL FIRE ACADEMY

". . . The Commission recommends that Congress establish a U. S. Fire Administration to provide a national focus for the nation's fire problem and to promote a comprehensive program with adequate funding to reduce life and property loss from fire."

With this recommendation, in the spring of 1973, the National Commission on Fire Prevention and Control actually took the first concrete step toward creating the U.S. Fire Administration.

In reality, however, factors that contributed to the creation of the Fire Administration were many and began in the late 1960s.

In 1967, three astronauts died in a fire aboard an Apollo spacecraft. That decade also saw disastrous fires as a result of riots that devastated many sections of this country's major cities; a fire in Montgomery, Alabama, took the lives of several national union officials, and a fire in a dormitory at Cornell University killed a faculty member and eight students.

While these tragic events were creating a climate for change among the public and government representatives, the fire service was creating its own climate for change with a major national conference, known as Wingspread I. Among the twelve "statements of national significance" that resulted from this conference was one critical to the concept of a federal focus on the nation's fire problem. That statement read, "The traditional concept that fire protection is strictly a responsibility of local governments must be reexamined."

The following year, the U.S. Congress passed the Fire Research and Safety Act of 1968. This legislation provided for many of the initiatives, including the collection of comprehensive fire data and fire safety education and training programs, that ultimately became the responsibility of the U.S. Fire Administration. However, the most significant section of this Act established a 20-member task force known as the National Commission on Fire Prevention and Control.

Unfortunately, the commission was not appointed until 1970 and went unfunded until 1971. Then, after nearly two years of research and hearings, the commission issued its report in May of 1973.

In addition to recommending the creation of the USFA, the commission made 90 other recommendations, covering the agency's structure, programs, funding, and cooperation with other government departments and private organizations.

The primary result of the commission's report, however, was the writing and enactment into law of the Federal Fire Prevention and Control Act of 1974 (P.L. 93–498).

The law created the National Fire Prevention and Control Administration and placed it in the Department of Commerce. The agency would be run by a presidentially appointed administrator who would report to the Secretary of Commerce. The National Fire Data Center and the National Academy of Fire Prevention and Control were specific parts of the Fire Administration, with the Academy Superintendent, though appointed by the President, reporting to the Fire Administrator.

The Fire Administration, among other responsibilities, was charged with:

- Conducting a major national program of public fire safety education.
- Assisting with the improvement of fire service training and education programs of all kinds at all levels.
- Promoting the drafting of master plans for fire prevention and control by local and state governments.
- Studying and assisting with the improvement of fire service management.

Accomplishments and Problems

While the Fire Administration's accomplishments are many and significant, the agency has been beset by a seemingly endless parade of administrative problems that many believe have hindered the USFA's effectiveness.

Problems

A complete explanation of the Fire Administration's difficulties is simply not possible in this brief overview. However, it is possible to fit most of the problems into the following categories:

- Lack of management continuity
- Numerous reorganizations and reprioritizations
- Several relocations of its headquarters
- An instability of the federal commitment (funding and authority)

During the first eight years of its existence, USFA had a total of nine acting and permanent administrators. In that same time frame, the Fire Academy had six acting and permanent superintendents.

Fortunately, management stability and continuity of leadership have come to both agencies. During the past five years, USFA has had just one Administrator and the academy has had only two superintendents.

The next two categories—reorganizations, reprioritizations and relocations—have been intertwined throughout the history of the Fire Administration.

When its work began in 1976, the Fire Administration developed a five-year plan that was to carry the agency through 1980. The plan divided the USFA into five program areas: general administration, Fire Academy, Public Education Office, Fire Data Center, and the Fire Safety and Research Office.

As early as 1977, a program to restructure and reorganize various federal agencies began to cast a shadow on that five-year plan. The Fire Administration was one of the agencies targeted for reorganization and, in 1979, the USFA was moved into the newly formed Federal Emergency Management Administration (FEMA).

The transition period was a time of instability and uncertainty for the Fire Administration, causing an interruption in the continuity of the USFA's programs and changes in personnel at the top of the Fire Administration and Fire Academy.

The transition also caused the development of a new five-year plan (1979–1983) with an altered structure and different program areas. The agency was divided into three divisions: Fire Academy, Office of Planning and Education, and Fire Data Center.

The program areas were:

- Arson prevention and control
- Residential fire safety
- Fire fighter health and safety
- Cost-effective fire protection management
- Fire incident command
- Emergency medical services
- First response to natural disasters
- First response to nuclear disasters
- First response to hazardous materials incidents

Changes in management and structure continued through 1980 and 1981, resulting in a major reorganization of FEMA in 1981. FEMA's new emphasis was to be on assistance to state and local governments, and this was to have a major impact on the Fire Administration and Fire Academy.

The Fire Administration was realigned into four divisions: Administration, Fire Data Center, Office of Fire Protection Management (formerly Office of Public Education), and Office of Fire Protection Engineering and Technology. For the first time, the USFA and NFA budgets were separated, with the Fire Academy budget falling under FEMA's Office of Training and Education.

After reviewing various suggested locations, The National Fire Academy was located on the site formerly occupied by St. Joseph's College in Emmitsburg, Maryland. The U.S. Fire Administration moved into the new FEMA Headquarters in Washington, DC.

This structure for the USFA lasted less than two years. In late 1982 and early 1983, the decision was made to move the Fire Administration to the National Emergency Training Center (NETC) campus in Emmitsburg and to reorganize again. The USFA was divided into four offices: Fire Policy and Coordination; Fire Fighter Health and Safety; Fire Prevention and Arson Control; and Fire Data and Analysis.

This instability has continued to the present. During the Higher Education Project Conference (1999), it was stated that plans were announced to move the Fire Administration to FEMA headquarters in Washington, DC. However, those plans were recently cancelled.

Constant reorganizations, reprioritizations, and relocations over the years have caused the loss of experienced personnel, a lack of continuity in the Fire Administration's programs, and a general uncertainty about the direction and priorities of the agency.

Funding for the federal fire programs has never lived up to initial expectations. The National Commission on Fire Prevention and Control originally recommended a budget of $125 million. Subsequent annual funding levels have not come close to matching this original, recommended budget level. In addition, shifting federal priorities and attempts to either eliminate or substantially reduce funding have caused significant fluctuations in the yearly budgets of the federal fire programs. These factors, simply stated, have only added to the uncertainty and instability.

Accomplishments

Despite the problems encountered by the federal fire programs during their existence, those agencies have a long list of accomplishments that have positively impacted the nation's fire problem at both state and local levels.

Arson Prevention Control

- Preparation of a report to Congress on the "Federal Role in Arson Prevention and Control."
- Development of the Arson Information Management System to identify early warning patterns.
- Development of an Arson Task Force Assistance Program to provide technical assistance and financial support to states and localities.
- Creation of an Arson Resource Center to serve as a national reference center.
- Development of a program to help local fire departments deal with the problem of juvenile fire-setters.
- Initiation of a three-phase study to identify the arson problem in rural areas of the United States.

Data Collection and Analysis

- Development and implementation of the National Fire Incident Reporting System (NFIRS).
- Publication of *Fire in the United States*, a detailed analysis of fire death, injury, and loss statistics.

Fire Department Management

- Development of a nationwide program to encourage master planning, including a report to Congress and publication of the *Urban Guide for Fire Prevention and Control Master Planning*.
- Initiation of a program to assist communities and fire departments in managing the entry of qualified women into the fire services.

- Preparation and publication of the *Report to Congress on the Effectiveness of Sprinklers and Detectors.*

Fire Fighter Health and Safety

- Publication and dissemination of an analysis of the causes of accidental death among fire fighters.
- Funding of Project FIRES, a program to develop state-of-the-art protective clothing for fire fighters.

Life Safety

- Completion of a review of state and local codes that require smoke detectors.
- Completion of a preliminary study of the impact of residential fire protection systems in actual fire deaths.
- Promotion of the use of smoke detectors through a nationwide smoke detector awareness campaign.
- Distribution of five million copies of *Smoke Detectors Save Lives* and *Wake up! Smoke Detectors Can Save Your Life.*
- Initiation of a program to train 9,000 local smoke detector specialists nationwide.
- Conversion of a fire safety evaluation system for hospitals and nursing homes into a training package for state and health care inspectors.
- Completion of mobile-home sprinkler tests to demonstrate the effectiveness of sprinklers in reducing fire hazards in mobile homes.
- Development and implementation of a number of programs and projects designed to promote the use of automatic detection and sprinkler systems throughout the country.

National Fire Academy

- Completion of a survey to assess the status of education and training for the nation's fire service.
- Development of the Open University, which will allow students to participate in accredited programs through independent self-study.
- Training of an estimated 350,000 students through the Academy Field and Resident Divisions and Train-the-Trainer programs.

Policy and Coordination

- Publication of the *Microcomputer Resource Directory for Fire Departments.*
- Publication of the *EMS Resource Directory.*
- Broadcasting of a series of teleconferences on a variety of topics to nationwide audiences.
- Establishment of the Integrated Emergency Management System (IEMS) to bring all elements of a local jurisdiction together in the disaster planning process.

Public Fire Safety Education

- Development of experimental public fire safety education programs for local, state, and federal agencies.
- Development of a five-step system for community fire education planning for use by local fire educators.
- Creation of the "Home Safety Surveys in Rural Areas" program.
- Sponsorship of the Children's Television Workshop's production of fire safety programs for "Sesame Street."
- Sponsorship of "Operation Dixieland" in cooperation with the state of Arkansas to impact Arkansas' fire death rate—the fourth highest in the nation.

APPENDICES

CONTENTS

APPENDIX 1

The *National Electric Code®* (NEC®) is the most widely used code in the world. It is published by the National Fire Protection Association. The NEC® is acknowledged to be the most widely adopted code of standard practices in the United States. It is also used extensively outside of the United States, particularly where American-made equipment is installed.

Chapter 1

Article 110 Requirements for Electrical Installation

Chapter 4

Equipment for General Use

Article 422 Appliances
Article 427 Fixed Electric Heating Equipment for Pipelines and Vessels
Article 440 Air Conditioning and Refrigeration Equipment
Article 450 Transformers and Transformer Vaults

Chapter 5

Special Occupancies

Article 500 Hazardous Locations
Article 511 Commercial Garages; Repair and Storage
Article 517 Health Care Facilities
Article 518 Places of Assembly
Article 530 Motion Picture and Television Studios
Article 545 Manufacturing Buildings
Article 547 Agricultural Buildings
Article 550 Mobile Homes
Article 551 Recreational Vehicles and Parks
Article 553 Floating Buildings

Chapter 6

Special Equipment

Article 600 Electric Signs
Article 610 Cranes and Hoists
Article 620 Elevators, Dumbwaiters, Escalators, Moving Walks, Wheelchair Lifts, and Stairways
Article 630 Electric Welders
Article 680 Swimming Pools/Fountains

Chapter 7

Special Conditions

Article 700 Emergency Systems
Article 760 Fire Protective Signaling Systems

Chapter 8
Communication Systems
Article 800 Communication Circuit
Article 810 Radio and Television Equipment

Additional National Fire Protection Association References

Automotive and Marine Server Code
Building Construction for the Fire Service
Emergency Management of Hazardous Materials Incidents
Fire Protection Guide to Hazardous Materials
Fire Protection Handbook
Fire Pump Handbook
Flammable and Combustible Liquids Code and Handbook
Haz-Mats Response Standards and Handbook
Industrial Fire Hazards Handbook
Learn Not to Burn Curriculum (3 levels for schools)
Life Safety Code and Handbook
LP-Gas Code and Handbook
Management in the Fire Service
National Fire Alarm Code and Handbook
National Fire Codes
National Fuel Gas Code and Handbook
NFPA 25–Standard for the Inspection, Testing, and Maintenance of Water-Based Fire Protection Systems
NFPA 52–Standard for Compressed Natural Gas (CNG) Vehicular Fuel Systems
NFPA 79–Electrical Standard for Industrial Machinery
NFPA 90A–Standard for the Installation of Air Conditioning and Ventilating Systems
NFPA 96–Standard for Ventilation Control and Fire Protection of Commercial Cooking Operations
NFPA 99–Health Care Facilities Standard and Handbook Set
NFPA 99C–Standard on Gas and Vacuum Systems
NFPA 231C–Standard for Rack Storage of Materials
NFPA 921–Guide for Fire and Explosion Investigations
NFPA Inspection Manual (7th Edition)
Occupational Health and Safety Standards Handbook
Principles of Fire Protection Chemistry
Truck Company Fireground Operations

APPENDIX 2

OSHA – 1997

Occupational Safety and Health Standards For General Industry (29 CFR-PART 1910)

Subpart L – Fire Protection

1910.156	Fire Brigades
1910.157	Portable Fire Extinguishers
1910.158	Standpipe and Hose Systems
1910.159	Automatic Sprinkler Systems
1910.160	Fixed Extinguishing Systems, General
1910.161	Fixed Extinguishing Systems, Dry Chemical
1910.162	Fixed Extinguishing Systems, Gaseous Agent
1910.163	Fixed Extinguishing Systems, Water Spray and Foam
1910.164	Fire Detection Systems
1910.165	Employee Alarm Systems (Also see: Appendices A–E to Subpart L)

The table on the next page contains a listing of those current national consensus standards which contain information and guidelines that could be considered acceptable in complying with requirements in the specific sections of Subpart L.

NFPA No. 10, Portable Fire Extinguishers
NFPA No. 11, Low Expansion Foam
NFPA No. 11A, Medium and High Expansion Foam Systems
NFPA No. 12, Carbon Dioxide Extinguishing Systems
NFPA No. 13, Installation of Sprinkler Systems
NFPA No. 13E, Fire Department Operations in Properties Protected by Sprinkler and Standpipe Systems
NFPA No. 15, Water Spray Fixed Systems
NFPA No. 16, Foam Water Sprinkler Systems and Foam Water Spray Systems
NFPA No. 17, Dry Chemical Extinguishing Systems
NFPA No. 18, Wetting Agents
NFPA No. 20, Installation of Stationary Pumps
NFPA No. 22, Water Tanks for Private Fire Protection
NFPA No. 24, Installation of Private Service Mains and Their Appurtenances
NFPA No. 69, Explosion Prevention Systems
NFPA No. 72, National Fire Alarm Code
NFPA No. 101, Life Safety Code
NFPA No. 1142, Water Supplies for Suburban and Rural Fire Fighting
NFPA No. 1963, Fire Hose Connections

Note: NFPA Standards are available from the National Fire Protection Association, 1 Batterymarch Park, Quincy, MA 02269.

APPENDIX 3

ADDRESSES

American Insurance Association, 85 John Street, New York, NY 12202
American Mutual Insurance Alliance, 20 N. Wacker Drive, Chicago, IL 60606
American National Standards Institute, Inc., 1430 Broadway, New York, NY 10018
American Society for Testing and Materials, 1916 Race Street, Philadelphia, PA 19103
Bureau of Explosives, 1920 L Street N.W., Washington, D.C. 20234
Business Publishers, Inc., 8737 Colesville Road, Silver Spring, MD 20910-3928
Division of Forest Fire Research, Washington, D.C. 10402
Factory Mutual System, 1151 Boston-Providence Turnpike, Norwood, MA 02062
Federal Fire Council, 19th and F Streets, N.W., Washington, D.C. 20405
FEMA, Emergency Management Institute, 16825 South Seton Avenue, Emmitsburg, MD 21717
Fire Marshals Association of North America, 1 Batterymarch Park, Box 9101, Quincy, MA 02269
Improved Risk Mutuals, 15 North Broadway, White Plains, NY 10701
Industrial Risk Insurers, 85 Woodland Street, Hartford, CT 06605
International Association of Fire Chiefs, 4025 Fair Ridge Dr., Fairfax, VA 22033
National Bureau of Standards, Washington, D.C. 20234
National Fire Protection Association, 1 Batterymarch Park, Box 9101, Quincy, MA 02269
National Safety Council, 1121 Spring Lake Drive, Itasca, IL 60143
Office of Admissions & Registration, National Emergency Training Center Telephone: (301) 447-6771 or (800) 638-9600 (Toll Free)
The Johnson Foundation, Racine, Wisconsin
U.S. Department of Agriculture Forest Service
U.S. Department of Commerce
Underwriters Laboratories, Inc., 333 Pfingsten Road, Northbrook, IL 60062-2096
United States Fire Administration, National Emergency Center, Washington, D.C. 20005
Wingspread Conference on Fire Service Administration, Education and Research, Washington, D.C. 20036

APPENDIX 4

FIRE CODES AND STANDARDS

THE EMERGENCE OF CODES AND STANDARDS

Greek and Roman law began to incorporate requirements for height, size, setbacks, and other features of buildings.These first construction requirements date back more than 5,000 years and had very limited application.

We know that in 1631 the Governor of Massachusetts issued an order banning thatched roofs to prevent fires from spreading from house to house. And the development of cities like New York, Boston, and Philadelphia eventually created a need for building codes. The first truly North American code was issued in New York in the late 1850s. By the end of the nineteenth century many cities had adopted building codes.

Modern building codes in the United States were prompted by disastrous conflagrations and earthquakes that occurred at the turn of the twentieth century. They dealt mainly with structural safety under fire and earthquake conditions. Since then, codes have grown into documents prescribing minimum requirements for structural stability, fire resistance, means of egress, sanitation, lighting, ventilation, and built-in safety equipment. More then 50 percent of a modern building code usually refers in some way or another to fire protection.

PUBLISHED CODES

The National Board of Fire Underwriters published the first model building codes in 1905. This organization, later called the American Insurance Association (AIA), distributed model codes until late in the last century. The model code theory suggests that a group of experts, with provisions for input from a broad spectrum of developers, can create a code that is a model for all jurisdictions to follow. Model codes make it easier for architects and other development professionals to work in more than one jurisdiction. These model codes deal with mechanical, plumbing, and fire prevention requirements.

Between 1915 and 1940 three major building code development organizations emerged. The Building Officials and Code Administrators International (BOCA) was established in 1915, while the International Conference of Building Officials (ICBO) and the Southern Building Code Congress International (SBCCI) formed later.

Building Officials and Code Administrators International (BOCA)

BOCA originally published the basic codes and AIA published the national codes. BOCA published the first edition of the *Basic Building Code* (BBC) in 1950. When AIA stopped publishing, BOCA started using the title *National Building Code*. Each of these codes is revised annually and reprinted every three years. The BOCA basic/national codes are used primarily in the Northeast, Midwest, and Mid-Atlantic. Changes approved between reprintings are published in supplements. BOCA maintains a technical staff in Country Club Hills, Illinois.

International Conference of Building Officials (ICBO)

ICBO was formed in 1922 and first published the *Uniform Building Code* (UBC) in 1927. Uniform codes are used in the West, Midwest, and Southwest, and have been adopted in municipalities as far east as Michigan.

The conference also publishes a *Uniform Mechanical Code* and *Uniform Plumbing Code* in association with the International Association of Plumbing and Mechanical Officials, and a *Uniform Fire Code* in association with the Western Fire Chiefs Association. Code changes are made each year, and amended versions of the codes are published every three years. Supplements containing any alterations are published between each major reprinting.

Southern Building Code Congress International (SBCCI)

The SBCCI was organized in 1940 and has published the *Standard Codes* since 1945. These codes are used primarily in the Southeast and Southwest.

Like the other two model code developers, SBCCI also publishes mechanical, plumbing, fire prevention, and gas codes. The codes are amended and reprinted every three years, with annual changes printed in supplements. SBCCI intends to promote standardization in building regulations and enforcement of those regulations. It maintains a technical staff headquartered in Birmingham, Alabama.

Note: Use of these codes is not restricted to a particular region; any developer around the country can use any one of the codes.

International Code Council (ICC)

The *International Building Code* is the product of the ICC. In the fall of 2001, the memberships of BOCA, ICBO, and SBCCI decided, in principle, to integrate these organizations within the International Code Council.

This integration stems from the memberships' resolution: "to work toward an ultimate goal of creating a single Model Code Organization." The public's best interest was at the heart of this integration decision. The leadership of the three organizations was encouraged to provide a single set of coordinated codes. It is this commitment to performance-based regulations that provides the United States with the best building codes in the world. Integration with ICC is another step toward offering effective and efficient regulations that meet public, government, and industry needs.

The three chief executive officers (CEOs), under the direction of their respective boards of directors, have been meeting to develop the organizational model, as well as transition and implementation plans for the single model code organization. To date, agreements have been reached on several key components.

Leadership among the three organizations confirms that this integration is acting on the members' resolutions and is proceeding as promptly as possible. Consolidating each of the operational functions means there are many legal issues and outside interests that have to be considered. The integration can only proceed at a pace that will result in the strongest and most efficient single organization. Based on the progress apparent today and the leadership commitments of the three organizations, a projected implementation date in 2003 is still realistic.

OTHER MAJOR BUILDING CODE DEVELOPMENT ORGANIZATIONS

There are two other important code development organizations, the National Fire Protection Association (NFPA) and the Council of American Building Officials (CABO).

National Fire Protection Association (NFPA)

This association's Standards Council has issued *NFPA 5000T—Building Construction and Safety Code T*. The association also publishes:

- *NFPA 1—Fire Prevention Code*. It addresses basic fire prevention requirements necessary to establish a reasonable level of fire safety and property protection from the hazards created by fire and explosion.
- *NFPA 70—National Electrical Code®*, which addresses proper installation of electrical systems and equipment.
- *NFPA 72—National Fire Alarm Code*, which sets minimum requirements for fire alarm systems and equipment.
- *NFPA 101—Life Safety Code*. This code provides minimum building design, construction, operation, and maintenance requirements needed to protect building occupants from the danger inherent in the effects of fire.
- More than 300 other codes and standards, manuals, and guides.

Council of American Building Officials (CABO)

CABO is an organization formed in 1972 by the three major model code groups. CABO's mission is to advance the model codes process on a national level and to work for uniform code regulations. The purpose of this council is to develop uniform code language to be included in each of the model codes. The two major accomplishments of CABO have been to organize the National Research Board and to publish the *One- and Two-Family Dwelling Code*.

The National Research Board also coordinates the research and evaluation programs of the three model code groups, eliminating the need for a manufacturer to work with three different organizations.

BUILDING AND FIRE CODES

Codes that deal specifically with construction of a building are part of a building code that is administered by a building department. A fire prevention code includes information on fire hazards in a building and is usually regulated by fire officials. Requirements for exits and fire extinguishing equipment generally are found in building codes, while the maintenance of such items is covered in fire prevention codes.

BENEFITS OF A CODE ENFORCEMENT SYSTEM

There are many benefits of a code enforcement system. Some major ones include:

- Lowering the threat of fire risk
- Reducing the incidence of fires and fire losses
- Improving life safety for the public
- Reducing hazards to firefighters and firefighting operations
- Controlling hazardous conditions
- Promoting a more stable community
- Maintaining a community's economic structure
- Providing community fire safety awareness
- Allowing code enforcement to be more easily implemented

IMPACT OF CODES AND STANDARDS

The implementation of codes and standards is seen in many aspects of day-to-day life. The placement of fire detectors and sprinklers, the design of building exits, and the installation of electrical wiring are just some of the areas influenced by fire codes and standards. The adoption of these codes and standards, along with increased public awareness of fire safety practices, has resulted in significant reductions in the loss of life and property damage due to the effects of fire. By continuing to use codes and standards that are readily accepted and followed, the world will be safer from fire and related hazards.

SUMMARY

Early building codes were concerned primarily with the prevention of building collapse. As civilization progressed and cities became crowded, regulations were formed to limit the types, number, and heights of buildings that could be constructed, and also to prevent the start and spread of fire in those buildings. As buildings and fire regulations developed throughout the United States, many were incorporated into law at the federal, state, and local levels of government.

Today's model building codes establish minimum requirements for construction and design, and fire protection codes and standards play an important part in community development.

REFERENCES

Federal Emergency Management Association, U.S. Fire Administration, and National Fire Administration. *Introduction to Fire Inspection Principles and Practices*. Washington, D.C.: Federal Emergency Management Association, U.S. Fire Administration, and National Fire Administration, February 1996.

National Fire Protection Association. *Principles of Fire Protection*, 6th ed. Quincy, MA: National Fire Protection Association, 1998.

GLOSSARY/FIRE TERMS

A

ACCESSIBILITY
Observations during inspection.

ACETALDEHYDE
Forms explosive peroxides when exposed to oxygen from the air.

ADAPTER
Fitting for connecting hose couplings with dissimilar threads with the same inside diameter.

ADVANCED EXTERIOR FIRE FIGHTING
Offensive fire fighting requiring the use of protective clothing, perhaps including self-contained breathing apparatus (SCBA), and performed outside a structure when the fire has progressed beyond the incipient stage.

AEROSOLS/FLAMMABLE LIQUIDS
Flammable aerosols or liquids present. Special hazards and their introduction into a general purpose warehouse can spell disaster. Flammable and combustible liquids are classified in *NFPA 30*, with Class IA representing the highest hazard and Class IIIB the lowest.

ALARMS-AUDIBLE NOTIFICATION
Bell, horn, chime, loudspeaker, or similar device that is actuated by a signal from an alarm-initiating device.

ALARM-INITIATING DEVICE
Mechanical or electrical device that sends an alarm signal to an alarm-indicating device. Alarm-indicating devices may be heat, smoke, or gas detectors; manual pull stations; or municipal fire alarm boxes. They may or may not be part of a fire suppression system.

ALARM SYSTEMS (FIRE DETECTION)
Equipment to detect the presence of fire and alert the occupants and notify the fire department.

AQUEOUS FILM-FORMING FOAM (AFFF)
Synthetic foam concentrate that, when combined with water, is a highly effective extinguishing and blanketing agent on hydrocarbon fuels.

AUTOMATIC ALARM
(a) Alarm automatically conveyed to local alarm bells and/or the fire station when actuated by heat, smoke, flame-sensing devices, or the water flow in a sprinkler system. (b) alarm boxes that automatically transmit a coded signal to the fire station to give the location of the alarm box.

AUTOMATIC SPRINKLER SYSTEM
An automatic sprinkler system in a facility provides a degree of safety that is incorporated into the interior finish requirements.

AUTOMATIC SPRINKLER VALVES
They must be open, and sprinkler standardized connections must be capped, free of debris, and accessible.

AUTOMATIC WET SYSTEM
There are five types of systems. The most common is an automatic wet system.

B

BLEVE
A boiling liquid-expanding vapor explosion. Bleve's are covered in more detail in the Fire Protection Handbook.

BOOSTER HOSE
Fabric-reinforced, rubber-covered, rubber-lined hose. Booster hose is generally carried on apparatus on a reel and is used for the initial attack and extinguishment of incipient and smoldering fires.

BOILING POINT
This is the temperature at which a liquid will boil, given in either °C or °F. The lower the boiling point, the more volatile and hazardous the flammable liquid.

BRAZING
A welding process in which the base metals are heated but not melted. Brazing can be accomplished with almost any fuel gas in combination with oxygen or air.

BUNKER EQUIPMENT
Each firefighter must be provided with protective equipment commonly referred to as "turnout gear" including helmet, coat, boots, trousers, and gloves, as required by OSHA 1910.156.

C

CARBON DIOXIDE (CO_2)
(a) Colorless, odorless gas that neither supports combustion nor burns; a waste product of aerobic metabolism. (b) Common extinguishing agent used in portable fire extinguishers.

CATALYST
Substance that modifies (usually increases) the rate of a chemical reaction without being consumed in the process.

CENTRAL ALARM SYSTEM
System that functions through a constantly attended location (central station) operated by an alarm company. Alarm signals from the protected property are received in the central station and are then retransmitted by trained personnel to the fire department alarm communications center.

CHARGE
To pressurize a fire hose or fire extinguisher.

CHEMICAL CHAIN REACTION
One of the four sides of the fire tetrahedron representing a process occurring during a fire: Vapor or gases are distilled from flammable materials during initial burning. Atoms and molecules are released from these vapors and combine with other radicals to form new compounds. These compounds are again disturbed by the heat releasing more atoms and radicals that again form new compounds and so on.

CLASS A FIRE
Fires involving ordinary combustibles such as wood, paper, cloth, rubber, and some plastics.

CLASS B FIRE
Fires of flammable and combustible liquids and gases such as gasoline, alcohol, kerosene, and propane.

CLASS C FIRE
Fires involving energized electrical equipment. Nonconductive extinguishing agents are necessary to extinguish Class C fires.

CLASS D FIRE
Fires of combustible metal such as magnesium, sodium, titanium, powered aluminum, potassium, and zirconium.

CODES
Documents written in a form and language appropriate for laws and ordinances. Their requirements are mandatory provisions using the word "shall." They set forth minimums to protect the health and safety of society and generally represent a compromise between

optimal safety and economic feasibility. Codes generally include administrative provisions, definitions, and requirements.

COMBINATION DETECTOR
Alarm-initiating device that is capable of detecting an abnormal condition by more than one means. The most common combination detector is the fixed-temperature/rate-of-rise heat detector.

COMBUSTION
Self-sustaining process of rapid oxidation of a fuel, which produces heat and light.

COMMAND
Act of directing, ordering, and/or controlling resources by virtue of explicit legal, agency, or delegated authority.

COMMAND POST
Command and control point where the incident commander and command staff function and where those in charge of emergency units report to be briefed on their respective assignments.

CONDUCTION
Transfer of heat by direct contact or through an intervening heat-conducting medium.

CONVECTION
Transfer of heat by the movement of fluids or gases, usually in an upward direction.

CORROSIVE
Can eat through metal and/or burn skin or eyes on contact.

COUPLING
Fitting permanently attached to the end of a hose; used to connect two hose lines together or a hose line to such devices as nozzles, appliances, discharge valves, or hydrants.

D

DAILY INSPECTION
In many facilities there are items that should be checked daily or at some other periodic intervals.

DECONTAMINATION - (DECON)
Removal of hazardous materials from anyone or anything that has been contaminated.

DEFLAGRATION
This is a combustion reaction that travels slower than the speed of sound while a "detonation" is a combustion that travels at or above the speed of sound.

DELUGE SPRINKLER SYSTEM
Fire protection sprinkler system in which the sprinkler heads are always open. The system is controlled by a valve that operates automatically by a thermostatically actuated device.

DISPENSING SYSTEMS
This generally involves the transfer of liquid from fixed piping systems, drums, or 5-gallon cans into smaller end-use containers.

DRILL
A simulated exercise conducted to practice and/or evaluate training already received; the process of skill maintenance.

DRY CHEMICAL
An ordinary dry chemical used primarily to extinguish flammable liquid fires. The most common include sodium or potassium bicarbonate, monoammonium phosphate, or potassium chloride.

DRY PIPE SPRINKLER SYSTEM
Fire protection sprinkler system that has air instead of water under pressure in its piping. Dry systems are often installed in areas subject to freezing. Refer to *NFPA 17 Standard for Dry Chemical Extinguishing Systems.*

DRY POWDER
Extinguishing agent suitable for use on combustible metal fires.

E

EGRESS
In a building, the exit access portion of the means of egress system generally comprises the majority of the floor area. Exit access includes all portions of a building through which an individual has to travel to reach an exit from any occupied spot in the building. Understanding the *Life Safety Code* requirements that apply to means of egress is of great importance. The distance a building occupant must travel to reach his/her nearest exit is termed the "travel distance."

EXPELLENT GAS
Any of a number of inert gases that are compressed and used to force extinguishing agents from a portable fire extinguisher. Nitrogen is the most commonly used expellent gas.

EXPOSURE
(a) Structure or separate part of the fire ground to which the fire could spread. (b) People, property, systems, or natural features that are or may be exposed to the harmful effects of a hazardous materials emergency.

EXTINGUISHING AGENT
Any substance used for the purpose of controlling or extinguishing a fire.

F

FEMALE COUPLING
Threaded swivel device on a hose or appliance made to receive a male coupling of the same thread and diameter.

FILM-FORMING FLUOROPROTEIN FOAM (FFFP)
Foam concentrate that is based on fluoroprotein foam technology with aqueous film-forming foam (AFFF) capabilities.

FIRE
Rapid oxidation of combustible materials accompanied by a release of energy in the form of heat and light.

FIRE ALARM SYSTEM
System of interconnected alarm-initiating and alarm-indicating devices designed to alert personnel to the existence of a fire in the protected premises. Alarm system may or may not be connected to a fire suppression system.

FIRE BEHAVIOR
Manner in which fuel ignites, flames develop, and heat and fire spread; sometimes used to refer to the characteristics of a particular fire.

FIRE BRIGADE
Employees within an industrial occupancy who are assigned at least basic fire fighting duties and responsibilities. The full-time occupation of brigade members may or may not involve fire suppression and related activities.

FIRE BRIGADE VEHICLES
Emergency response vehicles used by fire brigade personnel for fire suppression, rescue, or other specialized functions.

FIRE CONTROL ZONE
Uninvolved area immediately surrounding a fire, wide enough to protect fire brigade members and others from the adverse effects of the fire.

FIRE DEPARTMENT CONNECTION (FDC)
Point at which the fire department can connect into a sprinkler or standpipe system to boost the water flow in the system.

FIRE DETECTION DEVICES
Devices and connections installed in a building to detect heat, smoke, or flame.

FIRE DETECTION SYSTEM
System of detection, wiring, and supervisory equipment used for detecting fire or products of combustion and then signaling that these elements are present.

FIRE EXTINGUISHER
These portable held fire extinguishers are installed in many occupancies to give the occupants a means of fighting a fire manually. Portable fire extinguishers are not required by all occupancies and designed to fight incipient fires.

FIRE EXTINGUISHING SPECIAL AGENT SYSTEM
The most widely used special agent extinguishing systems are carbon dioxide, halogenated agents (soon to be eliminated), and dry chemicals. These systems initially are designed for a defined hazard. There are three basic methods used to apply extinguishing agents: total flooding system, local application systems, and hand hose lines.

FIRE-GAS DETECTOR
Device used to detect changes in the makeup of the atmosphere within a confined space as a result of combustion taking place within the space.

FIRE HOSE INSPECTION REQUIREMENTS
Fire hose found in commercial or industrial environments generally is intended either for occupant use in dealing with incipient fires or for use by trained firefighter s or brigade members in attacking a fire. The former is known as occupant use hose and the latter as an attack hose.

FIRE POINT
The point or temperature at which continuous combustion takes place.

FIRE PUMP
(a) Water pump used in private fire protection to provide water supply to install fire protection systems. (b) Water pump on a piece of fire apparatus.

FIRE STREAM
Stream of water or other water-based extinguishing agent after it leaves the fire hose and nozzle until it reaches the desired point.

FIXED MONITOR SYSTEM
Fire suppression system employing stationary master stream devices (monitors) in areas where large quantities of water or foam will be needed in the event of a fire.

FLAME DETECTORS
also called Light Detectors; these are used in some fire detection systems. There are two basic types: those that detect light in the ultraviolet wave spectrum (UV detectors) and those that detect light in the infrared wave spectrum (IR detectors).

FLAMMABLE
Capable of burning and producing flames.

FLASH POINT
The temperature at which a liquid gives off a gas or vapor sufficient to form an ignitable mixture with the air adjacent to the surface of the liquid.

FLASHOVER
Stage of a fire at which all surfaces and objects within a space have been heated to their ignition temperature and flame breaks out almost at once over the surface of all objects in space.

FOAM
Extinguishing agent formed by mixing a foam concentrate with water and aerating the solution for expansion; for use on Class A and Class B fires. Foam may be protein, synthetic, aqueous film forming, high expansion, or alcohol type.

FOG STREAM
Water stream of finely divided particles used for fire control.

FUEL
Flammable and combustible substances available for a fire to consume.

FUSIBLE LINK
Connecting link disk that fuses or melts when exposed to heat. These are utilized in

sprinkler heads, dampers, fire doors, and ventilators.

G

GALLONS PER MINUTE (GPM)
A measurement used by the fire services for the movement of water. Water used for fire protection must be segregated by values from the portable water that is delivered through the public water supply system. This is covered under *NFPA 24*.

GASES
The most common use for flammable gases is as fuel in gas-burning appliances and industrial heating equipment with air, and in cutting and welding processes with oxygen. Flammable gases are also used for medical purposes, principally with oxygen and nitrous oxide. Nearly all domestic and commercial gas appliances are covered by ANSI or UL Standards and are tested by the American Gas Association Laboratories, Underwriters Laboratories, and other testing laboratories. The release of gas from containers and piping does occur and this escape must be stopped.

GAS CYLINDERS
Cylinders that show signs of severe damage, corrosion, or fire exposure must not be used. The cylinder temperature should not be allowed to exceed 103° F. Always make sure that all labels and markings on the cylinder are legible and never use a cylinder whose contents are not known. Also, never store cylinders where they may be exposed to physical damage or to tampering.

GUIDES
Written by nationally recognized organizations, guides explain a code's or standard's written intent. They are advisory in nature and may give instructions, but do not contain mandatory provisions.

H

HALOGENATED AGENTS SYSTEMS
These agents or halogens have a number of unique fire protection qualities. Fire protection halogens currently are being phased out by the Montreal Protocol based on the fact they are linked to the destruction of the earth's stratospheric ozone layer.

HEAT DETECTORS
These detectors respond to the thermal energy signature of a fire and usually are located on or near the ceiling. All detectors should be tested periodically.

HAND LINES
A small hose line of 2 1/2 inches or less that can be handled without any mechanical assistance. Stations for 1 1/2-inch hose lines should be located throughout the premises so that all areas can be reached.

HYDRANTS
NFPA 25 calls for monthly inspections of any hose/hydrant houses to check accessibility, repair physical damage, and replace missing equipment is also an important task to cover. There is a specific checklist under *NFPA 25* that is specific to inspecting hydrants.

HEAT TRANSFER
This is the flow of heat from a hot area to a cold area. This flow may be accomplished by conduction, radiation, or convection.

HEATING SYSTEMS
A major consideration in the installation of any heat-producing appliance is the effect it would have on nearby combustibles, and installation clearances are very important, as is insulating combustible surfaces. Extensive information is covered in Chapter 4 of the *Fire Protection Handbook*, 17th Edition.

HOSE LINE
To transport water from a source of supply to a point of application, usually to suppress a fire.

HYDROSTATIC TESTING
According to NFPA, an "inspection" is required to give a reasonable amount of assurance that an extinguisher is available, fully charged, and is operable. "Hydrostatic testing" should be performed by personnel specifically trained for this task. The purpose of the test is to protect against the unexpected failure of the cylinder.

I

IGNITION SOURCES
All ignition sources should be controlled or eliminated in areas where flammable vapors could be present. The sources of ignition include open flames, heated surfaces, smoking, cutting and welding, frictional heat, static sparks, and radiant heat.

IGNITION TEMPERATURE
The lowest temperature required to start self-contained combustion, independent of an external heat source.

INCIPIENT FIRE FIGHTING
Activities involved in fighting incipient-stage fires inside or outside of buildings or other enclosed structures.

INCIPIENT PHASE
First phase of the burning process where the substance being oxidized is producing some heat, but the heat has not spread to other substances nearby. During this phase, the oxygen content of the air has not been significantly reduced.

INCIPIENT STAGE FIRE
Fire that is in the initial or beginning stage and that can be controlled or extinguished by portable fire extinguishers or small hose lines and without the need to wear protective clothing or breathing apparatus or to take evasive action such as crawling to avoid smoke.

INDICATING VALVE
Water main valve that visually shows the open or closed status of the valve.

INDUSTRIAL FIRE BRIGADE
Employees within an industrial facility who are assigned to respond to fires and other emergencies on that property.

INSPECTIONS
A written report should be prepared for each inspection by a fire inspector. The purpose of the report is to describe the property and its use, hazards, and fire protection without going into unnecessary details.

IONIZATION SMOKE DETECTORS
an ionization smoke detector contains a small amount of radioactive material. This ionizes the air in the sensing chamber, thus rendering it conductive and permitting a current flow through the air between two charged electrodes.

J

JERRICANS
The jerricans are used for liquids and are of metal or plastic packaging of rectangular or polygonal cross section. The maximum capacity of these cans is approximately 15.8 gallons (60 liters). Usually used in the transportation of hazardous materials such as antifreeze and a number of other specialty products.

JUNCTION BOXES
Junction switch and outlet boxes are used to protect the electrical equipment and connections that they house. The number of wires in a box must not exceed the number for which it was designed and should be equipped with the proper cover.

K

KNOCKOUTS
The various junction, switch, and outlet boxes are manufactured with "knockouts" that can be removed to allow the installation

of cable connectors and the entrance of the electrical cable.

L

LARGE HAND LINE
Fire hose/nozzle assembly capable of flowing up to 300 qpm.

LIFE SAFETY CODE
There are applicable *Life Safety Code* requirements for the appropriate occupancy classification. (NFPA, Volume 5, 1997). An example is that hotels over three stories high with guest rooms that open into corridors should have a fire alarm system, and seven or more stories high an annunciator panel to indicate the floor or area from which the alarm was transmitted are required.

LIQUID COMBUSTIBLES
The basic system for classifying liquids can be found in *NFPA 321, Standard on Basic Classification of Flammable and Combustible Liquids Code*. Flammable liquids have flash points below 100° F or more.

LOW EXPLOSIVE MATERIALS
Materials that produce deflagration or a low rate of reaction and pressure.

LP GAS
An LP-gas-air mixture, and natural gas also are used to fuel heat-producing devices. LP-gas vapors are heavier than air, and with such equipment in below-grade indoor locations it is critical for safety inspections.

M

MAINTENANCE
Poor maintenance and housekeeping practices are probably the principal reasons for fire problems. Improper storage of materials and poor maintenance on pumps, piping, and exhaust systems can make floors slippery, atmosphere dusty, and interfere with the proper operation of fire protection equipment.

MODEL CODE
A code created by a code development organization that has a special interest in a particular subject constitutes a model code; it can be adopted by any jurisdiction.

MOTORS
Motors and rotating machines can cause mechanical injury as well as a shock hazard and should be treated with caution. Many motors start automatically, so even a motor at rest should be treated as though it were running (lock-out procedure).

MUTUAL AID
This is a reciprocal agreement by organizations under a prearranged plan or contract that each will assist the other when needed for an emergency.

N

NONCOMBUSTIBLE MATERIALS
There are five fundamental types of construction. Type I is fire resistive-construction, *Type II is noncombustibles*, Type III is exterior protected combustibles, Type IV is heavy timber, and Type V is wood construction.

NONFLAMMABLE MEDICAL GAS SYSTEMS
Oxygen and nitrous oxide are used extensively in hospitals, dental offices, nursing homes, and other medical facilities for analgesic, anesthesia, and therapy. Because oxygen and nitrous oxide are nonflammable, their hazard as oxidizing agents is not readily recognized. The key fire safety precept for both gases is to keep fire away from them since their cylinders can be more dangerous than any other type of compressed gas cylinder during a disaster or fire.

O

OPEN FLAMES
Mostly used by restaurant owners to enhance atmosphere and with candles as food warmers. A source of ignition used is alcohol and generally is conducted very close to the restaurant patrons. Most jurisdictions allow this practice only when the restaurant is fully sprinkled.

OUTDOOR STORAGE
The storage of materials outdoors usually is limited to those used in large quantities and those not susceptible to damage by the weather. All outdoor storage should be structured so that it will not interfere with fire fighting access to and around building.

OFFENSIVE FIRE FIGHTING
Fire control activities intended to reduce the size of a fire and extinguish it.

OVENS
Ovens are defined as chambers used for baking, heating, or drying, or as chambers equipped to heat objects within the oven. Ovens usually operate at temperatures below 1400° F, although this does not always apply. Some coke ovens operate at temperatures above 2000° F.

OXIDATION
This is a chemical reaction in which oxygen combines with other substances. Explosions, fires, and rusting are examples of oxidation.

P

PALLETIZED/RACK STORAGE
Palletized storage consists of unit loads mounted on pallets that can be stacked on top of each other. Rack storage consists of a structural framework supporting unit loads, generally on pallets. *NFPA 231C* uses decision tables to specify when in-rack sprinklers are necessary and what type of in-rack sprinklers, either longitudinal or face, are needed.

PANIC HARDWARE
A locking assembly designed for panic exiting that unlocks from the inside when a release mechanism is pushed.

PHOTOELECTRIC CELL
Light-sensitive device used in some fire detectors. Cell initiates an alarm signal when light strikes it or is kept from striking it, depending upon the particular design.

POLAR SOLVENTS
Flammable liquids that have an attraction for water, much like a positive magnetic pole attracts a negative pole; examples include alcohols, ketones, and lacquers.

PRE-ACTION SPRINKLER SYSTEM
Type of automatic sprinkler system in which thermostatic devices charge the system with water before individual sprinkler heads are fused.

PREDISCHARGE ALARM
Alarm that sounds before a total flooding fire extinguishing system is about to discharge. This alarm gives occupants the opportunity to leave the area.

PYROLYSIS
Chemical decomposition caused by heat that generally results in the lowered ignition temperature of the materials

PROPELLANT
An explosive that normally functions by deflagration and normally is used for propulsion.

R

RADIOACTIVE MATERIALS
These materials are chemically identical to their non-radioactive counterpoints and pose identical chemical hazards. NFPA Standards recommend a safety analysis report, something major users and producers of radioactive

materials, such as Department of Energy plants licensed by the NRC, should prepare.

RADIATION
This is the transfer of heat through intervening space by infrared thermal waves. Also, energy from a radioactive source emitted in the form of waves or particles.

RADIANT ENERGY SENSING FIRE DEVICE
Designers specify flame and spark/embers detectors for sophisticated detection application.

RECOMMENDED PRACTICES
Advisory provisions that use the word "should" in the body of the text. This indicates a recommendation that is advised but not required. They are published by nationally recognized organizations and deal with maintenance and operational standards for the various systems required by the code.

S

SAFETY CONTAINER
Safety containers have a maximum capacity of 5 gallons and come with a spring-closing lid and spout cover so that the can will safely relieve internal pressure when subjected to fire exposure.

SIZE-UP
Ongoing evaluation of an emergency situation done mentally by the officer in charge, resulting in a plan of action that may be adjusted as the situation changes.

SMOKE
Visible products of combustion, which vary in color and density depending on the types of material burning and the amount of oxygen present.

SMOKE CONTROL
There are two recognized approaches to controlling smoke in buildings that make use of the air conditioning systems. The passive approach requires that fans be shut down and that smoke dampers in duct work be closed during a fire. In the active approach, the air conditioning system is utilized to exhaust the product of combustion to the outdoors to prevent smoke migration from the fire area.

SMOKE DETECTOR
Alarm-initiating device designed to actuate when visible or invisible products of combustion (other than fire gases) are present in the room or space where the unit is installed.

SOLID STREAM
Hose stream that stays together as a solid mass, as opposed to a fog or spray stream. A solid stream is produced by a smoothbore nozzle and should not be confused with a straight stream.

SOLUBILITY
Degree to which a solid, liquid, or gas dissolves in a solvent (usually water).

SPANNER WRENCH
Small tool primarily used to tighten or loosen hose couplings.

SPECIFIC GRAVITY
This is the weight of a solid or liquid substance, as compared to the weight of an equal volume of water. The specific gravity of water is (1.0). A liquid or solid with a specific gravity of less than (1.0) will float on water; if its specific gravity is more than (1.0), it will sink. If a liquid or solid is heavier than water, it can be extinguished by water.

Most flammable liquids are lighter than water; therefore, water cannot be used to extinguish the fire. An example is gasoline; burning gasoline floating on top of water can spread fire at a tremendous rate.

The specific gravities of the flammable and combustible liquids are very important factors in determining whether or not water can be used for extinguishing.

SPRINKLER
Water flow device in a sprinkler system. The sprinkler consists of a threaded nipple that connects to the water pipe, a discharge orifice, a heat-actuated plug that drops out when a certain temperature is reached, and a deflector that creates a stream pattern that is suitable for fire control. Also called Sprinkler Head.

SPRINKLER SYSTEM (AUTOMATIC)
An automatic sprinkler system in a facility provides a degree of safety that is incorporated into the interior finish requirements.

STANDARDS
Documents containing strongly recommended provisions and using the word "shall" to indicate requirements. Once adopted by a jurisdiction, standards become mandatory. They primarily provide technical, how-to details. A code might require a fire suppression system. A standard would list the requirements for design, construction, and installation of that system.

STANDPIPE HOSE
Single-jacket hose, lined or unlined, that is preconnected to a standpipe; used primarily by building occupants to mount a quick attack on an incipient fire.

STANDPIPE SYSTEM
Wet or dry system of pipes in a large single-story or multi-story building with fire hose outlets connected to them. The system is used to provide for quick deployment of hose lines during fire fighting operations.

STRAIGHT STREAM
The most compact discharge pattern that a fog nozzle can produce; similar to but not as compact as a solid stream.

SUPERVISED CIRCUIT
Alarm circuit on which a minute electrical current is constantly flowing. When this current is shorted or interrupted, an alarm or trouble signal is initiated.

SUPPLY HOSE
Hose between the water source and the attack pumper, laid to provide large volumes of water at low pressure. Also called Relay-Supply Hose or Feeder Line.

SUPPRESSION SYSTEM
System designed to act directly upon the hazard to mitigate or eliminate it, not simply to detect its presence and/or initiate an alarm.

T

THERMAL LAYERING
(a) Tendency of gases to form into layers according to temperature. (b) Characteristic of burning in a confined space, with the hottest air being found at the ceiling and the coolest air at floor level.

THREADED COUPLING
Male or female coupling with a spiral thread.

TOTAL FLOODING SYSTEM
Fixed, special agent fire suppression system that is designed to flood an entire room or space with agent to extinguish a fire. Halon and carbon dioxide are the two most common agents used for this purpose.

U

UNIT HEATERS
Unit heaters are self-contained, automatically controlled, chimney or vent connected air heating appliances equipped with a fan for circulating air. They can be mounted on the floor or suspended, and they are equipped with limit controls.

V

VALVE
Mechanical device with a passageway that controls the flow of a liquid or gas.

VENTILATION
Systematic removal of heated air, smoke, or other airborne contaminants from a structure and their replacement with a supply of fresher air.

VOLATILE
(a) Changing into vapor quite readily at a fairly low temperature. (b) Tending to erupt into violence; explosive.

W

WATER FLOW ALARM
Alarm-initiating device actuated by the movement (flow) of water within a pipe or chamber; most common installation is in the main water supply pipe of a sprinkler system.

WATER SUPPLY
Any source of water available for use in fire fighting operations.

WATER THIEF
Any of a variety of hose appliances with one female inlet and two or more male outlets, at least one of which is the same size as the inlet.

WET PIPE SPRINKLER SYSTEM
Automatic sprinkler system in which the pipes are filled with water under pressure at all times.

WINDWARD SIDE
Side of the building the wind is striking; the side of direction from which the wind is blowing.

WYE
Hose appliance with one female inlet and two or more male outlets, usually smaller than the inlet. Outlets are also usually gated.

INDEX

THE AUTHOR

Dr. Daniel E. Della-Giustina is Professor in the Industrial Management Systems Engineering Department and the Safety and Environmental Management Program, College of Engineering and Mineral Resources, at West Virginia University, where he has been a faculty member for 25 years. Dr. Della-Giustina has a Ph.D. degree in Safety, Health and Higher Education from Michigan State University, and the Educational Specialist degree from Michigan State University with an emphasis in Health and Traffic Safety Administration. Additionally, he has Master of Arts and Bachelor of Arts degrees in Liberal Arts and Behavioral Sciences from American International College, Springfield, Massachusetts. He currently is a member of the Board of Trustees at American International College.

Dr. Della-Giustina, a professional member of American Society of Safety Engineers, has published over 100 articles in the discipline of safety, health, and environmental studies and has presented scholarly papers at numerous meetings and conferences at the national and international level. He has presented papers at the International Sports Medicine Conference in Brisbane, Australia; the 2nd International Conference on Emergency Planning and Disaster Management, Lancaster-UK; *Crime and Its Victims—International Research and Public Policy Issues*, (NATO Conference) Tuscany, Italy; and *Fitness and the Aging Driver*, Stockholm, Sweden. During the summer of 2001, he served as an adjudicator in crisis management with the Freeport Company in Indonesia. He is currently editor of *The Safety Forum*, published by the School and Community Safety Society of America, Reston, Virginia.

He is a former Administrator of the American Society of Safety Engineers' Public Sector Division. At the 1995 ASSE Professional Development Conference, in Orlando, Florida, he was presented the ASSE Divisions' "Safety Professional of the Year" Award. In 2000, at the 66th Annual Conference of the West Virginia Safety Council in Charleston, Dr. Della-Giustina was presented the "Safety Professional of the Year" award by Governor Cecil Underwood.

Dr. Della-Giustina has appeared as an expert witness in safety and health liability cases for the past 20 years in numerous cities throughout the United States. He has served as consultant to various public, volunteer, and industrial fire brigades in the areas of disaster preparedness and emergency systems. During the past five years, Dr. Della-Giustina has been a member of numerous committees with the American Society for Testing and Materials (ASTM).

In January 1998, he was appointed as a Principal Member of the National Fire Protection Association's Technical Committee on Industrial Fire Brigades' Professional Qualifications.

The American School and Community Safety Association has presented Dr. Della-Giustina its Scholar Award (twice), and its Presidential Citation as well. Dr. Della-Giustina's distinguished record as a researcher, author, teacher, and administrator has made him a national leader in the safety profession.

In September of 2001, at the Congress of the National Safety Council in Atlanta, Georgia, Dr. Della-Giustina was inducted into the Council's Safety and Health Hall of Fame International. The Hall of Fame is dedicated to recognizing leaders and pioneers for their innovative contributions to the safety, health, and environmental industry worldwide.